COOPERATIVE EVOLUTION

RECLAIMING DARWIN'S VISION

**CHRISTOPHER BRYANT
AND VALERIE A. BROWN**

Australian
National
University

PRESS

For Annie Bryant

who has had to put up with parasites and mitochondria at the dinner table for 58 years and, in spite of that, managed to pass on to our children a very healthy mitochondrial genome, with love.

And for Chris, Sarah, Elliot and Amon Brown

who are taking the next steps in cooperative evolution.

By mutual confidence and mutual aid,
Great deeds are done and great discoveries made.

– Homer

ANU PRESS

Published by ANU Press
The Australian National University
Acton ACT 2601, Australia
Email: anupress@anu.edu.au

Available to download for free at press.anu.edu.au

ISBN (print): 9781760464288
ISBN (online): 9781760464295

WorldCat (print): 1240754622
WorldCat (online): 1240754606

DOI: 10.22459/CE.2021

Cover design and layout by ANU Press

CONTENTS

ACKNOWLEDGEMENTS

We are most grateful to Emeritus Professor Sue Stocklmayer, Dr Hugh Tyndale-Biscoe, Dr Adrian Verrender and Professor Kim D Rainsford, who read the manuscript at various times and gave us detailed feedback. Their comments have greatly improved our work but its shortcomings are entirely our own responsibility. The late Professor Colin Groves was always available to answer questions and encourage us further.

Chris Bryant is particularly grateful to members of his family, whom he used relentlessly as sounding boards:

- Grandson Harry Rumble first read the earliest completed version and his enthusiasm for it was a great reassurance.
- Son Tim Bryant heartened him further by remarking, with raised eyebrows and a note of disbelief in his voice, that 'it was surprisingly readable!'
- Daughter-in-law Peta Swarbrick continued the process of encouragement by quoting his own work back at him during an argument.
- Grandson Ali Bryant pointed the pistol shrimp at him, as an unusual example of symbiosis.

The authors are grateful to ANU Press for a subsidy to support the publication of this book.

All figures were prepared for the authors by Vladimir Zokic, unless otherwise indicated.

GLOSSARY OF WORDS AND PHRASES

ancient cell	the original bacteria-like cells; prokaryotes
ATP	a molecule that contains usable energy
chloroplast	the green body (or organelle) in the cell of a plant, responsible for photosynthesis
entropy	in thermodynamics, a measure of disorder
epigenetics	the modern version of Lamarckism
Gaia	Lovelock's name for the self-maintaining and self-organising planet Earth
gene	the unit of heredity
gene expression	the translation of a genome to make an organism
genome	all the genes of a single organism
genus	a group of similar organisms; e.g. dogs belong to the genus *Canis*
homeostatic	a homeostatic system is one which, if its current steady state is disturbed, tends to adjust in such a way as to restore it
horizontal gene transfer	transfer of a gene from one individual to another of the same generation
Lamarckism	the inheritance of acquired characteristics
mEve	a hypothetical ancestress of all humans
mitochondrion (pl. -ria)	the organelle, once itself an ancient cell, responsible in modern cells for the generation of energy
modern cell	a symbiotic construct of three or four ancient cells; eukaryotes

organ	a group of tissues aggregated to perform a single function; e.g. the human liver
organelle	by analogy with organ, a structure within the cell that performs a specific function
Protista	single-celled plants and animals
species	the second order name of an individual animal or plant; e.g. dogs, *Canis*, belong to the species *familiaris*
symbiosis	the condition where two or more separate individuals are mutually dependent for their continued existence
tangled bank	Darwin's metaphor for the interdependence of living things
tissue	a group of cells each performing the same function within an organism
yAdam	a hypothetical ancestor of all humans

INTRODUCTION

A cooperative enterprise; how this book came to be written.

If you choose not to read this and move straight on to *In Homage to Darwin,* we will not be offended. The story, however, of how two scientific dinosaurs came to cooperate in writing about their shared world view, gleaned from a combined total of more than 150 years of independent study of biology, needs, we feel, some explanation.

Any two biologists can be relied on to arrive at different personal understandings of Charles Darwin's magnificent legacy. However, few were as far apart as the authors of this book when they first met in the 1970s. At that time, a split was appearing in the practice of science. Traditionalists were persevering with the reduction of whole systems into their constituent parts, an approach that had led to the triumphs of the decipherment of the genetic code and the new science of genomics. A different way of thinking was combining science and new social movements. Post-normal science was beginning to accept that, for complex issues such as planetary climate change and global food security, scientists needed to practise their art where facts were uncertain, values in dispute, stakes high and decisions urgent.

The authors met, 45 years ago, in the Department of Zoology at The Australian National University (ANU). Chris Bryant was then a reader, with a flourishing research group in parasite biochemistry. He had remained a reductionist, focusing his attention on the subcellular mechanisms of respiration in anaerobic organisms. Val Brown, having raised a family, was a mature-age PhD student working in the then-new field of holistic thinking as applied to the human sciences. She was already a fan of Lovelock's space-engendered view of the Earth as a self-maintaining and self-organising planetary system he named 'Gaia'.

They did not hit it off.

It is hardly surprising. They had begun at opposite ends, both of the world and of their discipline. Chris was born in North London in 1936. He was educated at an English public school. He first entered a biological laboratory in 1948 and instantly fell in love with the smell of solvents, the specimen cases and the microscopes. He came under the influence of Theodore Savory who, though a schoolmaster, was a major and well-published authority on spiders. He was also the author of a very reductionist, but well thought-of, book expressing these views (Savory 1936). After six years of exposure to his excellent teaching, Chris, like his mentor, was a confirmed reductionist. He felt that by studying the minutiae of organisms he would eventually come to understand the whole.

This view of life was not dispelled by his time as an undergraduate at London University. After completing a Master's degree and a PhD, in 1961, he met and married Anne, an Australian nurse. He decided to try his luck in Australia. He accepted a lectureship in zoology at ANU. Twenty-five years later he was appointed to its chair.

Val's first lesson in collective thinking occurred on leaving her conservative family and Anglican convent school for the University of Queensland. There, she found herself one of six women among 700 men and, despite distractions, obtained a combined zoology and botany degree. At this time, the influences on her thinking were the zoologist William Stephenson, well known for his work on the Great Barrier Reef, and the botanist Desmond Herbert, a biogeographer with a passionate interest in the subtropical rainforest. They were field biologists, and between them they cemented Val's love for interactive biological systems as they occurred in the 'real' world.

On graduating, she accepted a post of research officer, and so became the first woman scientist in the CSIRO in Brisbane. Sadly, it was then the rule for women in government employ to resign when they got married, and she had to leave. After raising three children, she returned to academia at The Australian National University's Department of Zoology to undertake a Master's degree that grew into a PhD on *Holism in the University Curriculum?* The question mark was important because she found that, while the curriculum professed to be holistic, it wasn't.

Neither followed the expected pathways in their original fields. Dissatisfied with the context-free subcellular reactions that he had been studying, Chris began to pay more attention to the host–parasite relationship in its entirety, moving much closer to a holistic view of life. He developed an interest in the adaptive relationship between the parasite and its host environment and the interdependence that existed between non-parasitic organisms. Moving even further from mainstream reductionism, he relinquished his chair in 1996 and then, as professor emeritus, moved on to help establish the National Centre for the Public Awareness of Science at ANU, of which he was the first director.

On a converging track, Val applied the findings of her PhD to practical ways of integrating social and physical sciences in health, education, environmental management and government. As a community service, she acted as consumer representative on major national decision-making bodies in these same fields. In time, this developed into a national research centre, the Local Sustainability Project at ANU. Val went on to become professor emerita in Environmental Health at Western Sydney University, and then took the Local Sustainability Project back to ANU in 2003.

After Chris and Val crossed swords in the 1970s, their paths did not merge again until 2014. Then they found themselves sitting next to each other at a celebration of the life of a mutual colleague. They started a conversation and discovered that there were now many common elements in their scientific thinking and reading, and that, over the years, their ideas had converged towards the concept of a wholly integrated, planetary Gaia-like biosphere that included human society. It was as if they had been part of a landscape in which they travelled separately through a forest, along a track leading towards the distant horizon. Val had always had her head in the air and her eyes fixed on the tree canopy while Chris's attention had been concentrated on the detail of the understorey. They agreed that it was high time to see if their world images were congruent – whether they were indeed travelling in the same forest.

Not only did they find that their professional experiences had led them to the same conclusions, they had arrived at a mutual interest in the interactions between biophysical and social evolution, and a deep admiration for Charles Darwin. The conversation moved from deploring the commonly held but unscientific interpretation of Darwin's work as nature 'red in tooth and claw' when he himself preferred the idea of an all-connecting natural system – a 'tangled bank'. So began a series of conversations that led to this book.

The outcome? Two people coming from opposite ends of a scientific tradition have presumed to retell the story of the evolution of the integrated planet – and with as little technical jargon as possible.

1

IN HOMAGE TO DARWIN

In which we explore Darwin's vision of a cooperative evolution.

In writing this book, we were overwhelmed by the recent deluge of new information about the evolutionary history of our own genus, *Homo*. We consulted an expert colleague, Professor Colin Groves, an authority on human evolution. He asked what our book was to be about. We answered that, in the light of modern discoveries, our aim was to put forward a very different view of Darwin's own, personal approach to evolution. The view of evolution, not Darwin's, usually portrayed in the popular press is of nature, red in tooth and claw, struggling for survival. It gives the impression that Darwin had thought that every living thing was at war with every other living thing. This is incorrect and we wanted to explore the complex dynamic cooperation among all life forms, including ourselves, that Darwin considered a constant theme of evolution.

'Good,' said Professor Groves, 'it's about time someone did'.

We were delighted to have the approval of a respected colleague. It certainly appears that this story is ready to be told.

At a time when the world is being driven by division and confrontation, it is heartening to look more closely at what Darwin has to say. In the last paragraph of his great work, *On the Origin of Species* (1859), Darwin makes an impassioned plea in a way that any man or woman familiar with the countryside could understand:

> It is interesting to contemplate a tangled bank, clothed with many plants of many kinds, with birds singing on the bushes, with various insects flitting about, and with worms crawling through

the damp earth, and to reflect that these elaborately constructed forms, so different from each other, and dependent upon each other in so complex a manner, have all been produced by laws acting around us … There is grandeur in this view of life.

Later, in another great work, *The Descent of Man* (1871), he painted humans into this picture.

There is indeed grandeur in 'this view of life', of an evolutionary process, driven by ever-changing environments and the interactions and collaborations among life forms. The tangled bank is a microcosm, the world writ small. But what is true for tangled banks is true for a single cell or an elephant, for forests and prairies, for tundras and alpine pastures, for farms and cities, for communities and nations. All are subject to the same laws and all interact with adjacent ecological systems until the whole of life is a single operational entity. Humans may have added their social dimensions but the interaction with biophysical laws still applies. Environmental changes have unprecedented feedback systems which we do not yet understand.

The science is clear. The carbon of fossil trees and plants that lived 300 million years ago is now being burned to release carbon dioxide, which is changing the dynamics of the atmosphere. Modern forests, which extract carbon dioxide from the atmosphere and return oxygen in an endless cycle, are being felled or burned at an ever-increasing rate. The temperature is rising. Permafrosts, no longer permanent, are warming to release their greenhouse gases. The poles are melting, releasing floods of fossil water that are affecting ocean currents and influencing the weather (see, for example, Eggleton 2012).

The fabled butterfly that flaps its wings on one side of the globe and causes atmospheric disturbance on the other is hard at work. World ecology is changing and the existence of many plants and animals is threatened by human activities. Human populations are moving around the planet, whether willing or unwilling. Human communities differ on whether to curb industrial processes that bring affluence to some and misery to others.

It is an era of fragmentation – of knowledge, of interests, of politics and, most of all, of world systems. It is more obvious than ever before that humanity, by its ingenuity and ever-increasing population, is largely responsible for these changes. As we write, the need for collaboration between nations is so strong that it surely demands a new world order, one where an understanding of Darwin's ideas can be as great a service to the twenty-first century as it was to his own.

Charles Darwin was far from the single-minded, conservative patriarch suggested by his public image. Rather, he was an open-minded and courageous risk-taker to the end. At a time when most people believed that the world as a whole was unchangeable, Darwin was brave enough to question that belief. Writing to his friend Joseph Hooker in January 1844, Darwin revealed the extent of his inner conflict between Christian fundamentalism and the open-ended evolution of species. He wrote, 'I am almost convinced (quite contrary to the position I started with) species are not (it is like confessing a murder) immutable'. Fifteen years later, his own conflict resolved, he allowed *On the Origin of Species* to be published and went on to profess an even greater heresy: human beings were also subject to evolutionary change.

Darwin's insights provide the means for us to understand ourselves, how we originated on this planet and even to imagine where we might be going next. He changed the world we live in: away from one specially created by a deity for humans, who were the only intelligent beings, giving them a right to exploit all of creation. We are still coming to terms with a living world whose dynamic systems choose their own direction, and in which all manifestations of life display their own form of intelligence, where humans are only one species among many.

In the years since Darwin, his view of evolution has been distorted by his successors. Even while he was alive, it was variously rendered as 'nature red in tooth and claw', 'survival of the fittest' and 'the struggle for existence', with the emphasis on aggression between species and between individuals of the same species. Of course, individuals of the same species do compete with one another for access to resources – but it is a passive competition, brought about by living in the same geographical space. Generally, species are not at war with each other – even predators are not at war with their prey; they are focused on their own survival. Aggression, however, is a human concept applied erroneously to other living things. Among plants, some are designated as 'weeds'. This too is a human construct. Weeds are not aggressive; they are simply plants optimising their own survival by getting their share of available resources. Unfortunately, humans see them as a threat to the fruits of their labours on farms and gardens that are so far out of ecological equilibrium that they cannot survive without constant attention. And sadly, humans seem to be the only animals to show purposeful aggression to others of their own species, to the point of systematically killing them. Among animals, lethal aggression is rare, and death, as occasionally occurs when two stags fight over a female, is rare 'collateral damage'. Darwin was, of course, aware of all this at the same time he was writing about the tangled bank.

It is time, 160 years later, to return to a view of evolution as Darwin envisaged it. Looking beyond the microcosm of the tangled bank, we can now see it as a global system of cooperation and mutual aid. Sadly, even well-meaning disciples distorted Darwin's original message. They were Victorians, with minds that were conditioned by the concept of *Rule Britannia* and conquest. The catchphrase 'survival of the fittest', coined by the enthusiastic Herbert Spencer, contributed nothing but a convenient empty and self-referential tag. Who are the 'fittest'? They are those that survive. Who are the survivors? They are the fittest. The British Establishment loved it, for humans were obviously the fittest, as the only intelligent beings, and the British were the fittest of all. An alternative view, 'collaborate and flourish', is a far better description of the course of evolution.

Now imagine that you are a traveller in space and time. With the speed of thought, in an instant that is unhampered by Einstein's Relativity, you can traverse vast distances and cross aeons of time. On your journey through the galaxy, your attention is caught by an unremarkable solar system, out near the edge. It becomes focused on a small stony planet with an iron core, third from its sun, with an atmosphere of carbon and sulphur gases, ammonia and nitrogen but no oxygen. There is nothing remarkable in that.

Lightning is glowing in the upper atmosphere but then something else catches your eye. Under your mental microscope it appears to be almost alive – perhaps it is alive. It is the first cell on the surface of this ball of rock, warmed by the life processes going on within it, and by the sun's energy from without. As the millennia go by, cells come and go; these are the ancient cells, trillions of which persist to this day. Eventually, after enough changes, a new and larger cell, a modern cell, appears. It is the entry point for the remarkable evolutionary journey that Darwin has described and which is taking place around us.

Darwin was very much a product of his century; a young, upper-middle-class man, with an extensive, intellectual and influential family background. There is nothing much in his early years to indicate that his ideas would change the way in which humanity constructs its concept of the world. His career had begun to unfold along conventional Victorian lines; a mediocre student with an interest in natural history – he was a keen 'beetler', a collector of insects – with probably a good career beckoning in the church. He might reasonably have expected, with his connections, to finish up, one day, as a bishop, with time only to write the odd natural history letter to the Linnean Society.

What set Darwin apart from his particular cohort was the sense of adventure and inquiry that led him to accept the post of naturalist aboard the *Beagle*. He went off the Victorian rails and never regained them, for the voyage of the *Beagle* lasted five years. And it was a critical five years in the life of a young man, starting at his coming of age at 21. This is a period when he is normally being fitted into a productive life within his social set. In fact, Darwin was moulded into another shape entirely – a shape that was about as far away from Victorian convention as one could imagine. He was not, however, the first to consider the changes which created new species. His own grandfather, Erasmus Darwin, had already been laughed at for putting forward the same general principles.

Even before he embarked on the *Beagle* it was apparent that he possessed a great capacity for reason and objective analysis coupled with an ability to set out the logical steps in an argument. A trivial example – or perhaps not so trivial, in view of later events – is the letter in which he carefully lays out his father's objections to his accepting the offer of a post on the *Beagle* (Darwin 1887). His father had agreed to give his permission if Charles could persuade a much-admired uncle that it was a valuable opportunity. Charles drew up a list of the points that he had to refute:

1. Disreputable to my character as a Clergyman hereafter.
2. A wild scheme.
3. That they must have offered it to many others before me, the place of Naturalist.
4. And from its not being accepted there must be some serious objections to the vessel or expedition.
5. That I shall never settle down to a steady life hereafter.
6. That my accommodations will be most uncomfortable.
7. That you should consider it as again changing my profession.
8. That it would be a useless undertaking.

Darwin's careful advocacy prevailed. He obtained the support of the uncle and the world was changed.

To us, now, many of the objections raised by Darwin's father seem rather desperate, a father trying to dissuade a beloved son from undertaking a voyage that could easily end in disaster and death, while at the same time leaving these fears unsaid. Darwin treated each objection seriously, giving them equal weight and showing his capacity for the ordered and organised thinking that was so essential for his future work.

Convincing an uncle was not a particularly complex problem. Unravelling the processes that underlie the structure and distribution of coral reefs was. Simply stated, the problem was to understand how it is that coral forever grows upwards yet always remains at the same depth for optimum growth.

In the modern world view of plate tectonics and an ever-shifting Earth the answer seems obvious; the seabed must sink at the same rate that the coral grows and/or sea levels must rise. This was not so obvious in the mid-nineteenth century, given the state of knowledge then. His thoughts about coral reefs implied a connection between corals and a dynamic Earth that threatened the stability of the conservative world inhabited by Victorians. Darwin published his personal observations and the fruits of his subsequent researches in 1851, in a treatise entitled *Geological Observations on Coral Reefs, Volcanic Islands and on South America*. The section on the structure and distribution of coral reefs is a masterpiece of detailed scientific argument. In turn, he deals with atolls, barrier reefs and fringing reefs and brings them all together in an overarching hypothesis for their formation. Any doubter is buried under a mountain of facts and observations and is left with nowhere to go. This is the technique, compounded of great wisdom, organisation and thoroughness, that he used to such good effect in his great work, *On the Origin of Species*.

Darwin's other great attribute was to so enthuse others about his work that, if he asked for help, they felt privileged and eager to collaborate. In later years, he was beset by a mysterious illness, thought by some to have been due to stress brought on by his 'confessing a murder'. He found it difficult to leave Down House for more than a short time and made up for it by being a prolific correspondent. His letters are a delight to read, often light-hearted, always courteous, always generous, especially with praise for the ideas of others or in his gentle criticisms of them. In particular, he was always very generous and self-deprecating when dealing with Alfred Russel Wallace, the co-discoverer of natural selection and whose earlier work he admired. The feeling was reciprocated. Darwin would sometimes indulge his exasperation with those who failed to see the point of an argument, but only in letters to close friends in whom he had absolute confidence. In this way he sat at the centre point of a great network of information that was coming to him from admirers and co-workers everywhere on the globe. In many ways he reflected the great phenomenon he was himself illuminating – thriving at the centre of a great system of collaborative work.

Even in Darwin's own lifetime, while he was writing *On the Origin of Species*, and after – he lived for another 23 years – facts corroborating evolution came pouring in. For example, in 1856 Louis Pasteur showed that fermentation is caused by microorganisms – that is, that decay and the return of nutrients to the environment to complete the great cycle of life and death is a function of living things. The 'cell doctrine' of Virchow in 1858 proclaimed that all cells arise from existing cells and sounded the death knell for the belief in spontaneous generation. Mendel's famous experiments on inheritance were published in 1868 but not translated into English until 1900. It is a minor tragedy that Darwin died unaware of them, because he was already thinking about the idea of inheritable 'particles' that he called gemmules. Miescher found and named 'nucleic acids', the important component of Mendel's particles, in the nuclei of cells in 1869. The details of all these pioneers may be found in any good history of biology, such as that of Singer (1959), and the first 40 years of the twentieth century have been summarised by Huxley (1942). It was a further 10 years before the structure of nucleic acid (DNA) was worked out by Watson and Crick (1953).

Thus, in the first half of the twentieth century, support for 'Darwinism', as it was then called, came thick and fast. In 1902, chromosomes were identified as carriers of genetic information, and William Bateson invented the term 'genetics' to describe the new scientific discipline that developed as a consequence. Chromosomes were found to be made up of strings of individual genes (the particles of inheritance). Studies of gene behaviour led to mathematical genetics and at last a sound theoretical basis for Darwin's natural selection was established.

As the century unfolded, more and more information about the various processes and manifestations of Darwin's natural selection accumulated. In 1942, Julian Huxley brought this all together in one great book and

Neo-Darwinism was born. The vast majority of biologists all over the world were now evolutionists. When Oswald Avery, in 1944, showed that DNA carried the hereditary information in a bacterium, attention became focused on DNA and an almost unseemly rush to elucidate its structure ensued. In 1948 Erwin Chargaff showed that, in the structure of DNA, the proportions of the chemical bases guanine (G), cytosine (C), adenine (A) and thymine (T) were rigidly controlled in any organism. Without this knowledge, the publication in 1953 of the famous paper by Watson and Crick that established the helical structure of DNA could not have occurred.

The second half of the twentieth century saw many more scientists at work than ever before in the history of humanity; Darwinism became a flood of such proportions that it is impossible to identify an orderly progression through the years. Here are some highlights.

The DNA code of G, C, A and T base sequences has been worked out, and a new discipline, genomics, is providing seemingly unlimited insights and opportunities. The science of ecology has been put on a formal footing, and out of this has arisen the disciplines of social ecology and social evolution that are so important for understanding humanity. The old, formerly discredited, Lamarckian theory of the inheritance of acquired characteristics, which Darwin did not, in fact, dismiss, has emerged as the new science of epigenetics. It now transpires that changes in DNA can occur during an individual's lifetime and they can be inherited by offspring. Symbiosis, that we might call cooperative biology, is now recognised as a major driver of Darwinian evolution and is the subject of this book.

Darwin clearly recognised the source of the grandeur of his 'tangled bank'; it was the collaboration that connects all the changing forms of life on planet Earth. After Darwin, it is hard to imagine a time when the general belief was of a world that was created some 4,000 years ago and in which all forms of life had remained unchanged since.

The subtitle of *On the Origin of Species* is: 'by means of natural selection or the preservation of favoured races in the struggle for existence'.

While Darwin's writings have proved to be timeless, it is still necessary to take account of the changing times. The term 'races' here would have been interpreted as 'species' or 'families'. In his third chapter he explains that the environment in which living organisms exist is limited and finite.

Members of the same species depend on the same resources. When resources are short, they may be in competition with each other, while one of their survival techniques can be collaboration within and between species.

Darwin is at pains to emphasise that, allowing for this, 'the relation of organism to organism is the most important of all relations' (*Origin*). We can now add to this cell/cell relationships – in a developing embryo, for example many cells 'sacrifice' themselves for the common good by succumbing to a 'programmed cell death'.

The 'struggle' is perhaps best explained by a fable. Imagine that two people swimming in the sea get caught in a rip. Both will struggle to survive but they struggle against the current, not each other. If neither of them is a good swimmer, both may be drowned. However, if one is a good swimmer, that person may be able to save both themself and the other. At some later date the one who was saved may be in a position to drag someone else from a different danger. Interaction can influence which species characteristics are preserved and which are lost, due partly to chance, and partly to the changing state of the environment. Survival rests on connections among the participants and compatibility with the impersonal environment.

On its 'splash page', the Darwin Project (www.thedarwinproject.com), concerned primarily with human evolution, proclaims that 'in the *Descent of Man*, Charles Darwin wrote *only twice* of "survival of the fittest" – but *95 times about* love!' (italics in original). Love is an extreme form of interdependent cooperation displayed by humans and, in some form, perhaps as maternal care, by higher animals.

Pyotr Kropotkin, in his book *Mutual Aid*, describes his observations of wildlife in the harsh climate of Siberia and how he was struck by the many examples of positive interaction that it displayed. He noted that Darwin wrote in *The Descent of Man*:

> … how, in numberless animal societies, the struggle between separate individuals for the means of existence disappears, how struggle is replaced by co-operation, and how that substitution results in the development of intellectual and moral faculties which secure to the species the best conditions for survival.

Darwin *was* the cutting edge of his time and could not possibly have known what we now know. On the other hand, his recognition of the connections between all living things, between life forms and the environment, and between human beings and the rest of the world, has given us the understanding of the world we have today and the world we may expect tomorrow. He did not make the mistake of expecting the world to stay the same, as his work on coral reefs shows. Evolutionary pressures continue today, with the influence of human social activity triggering greater changes than ever before.

And this is what our book is about: the golden thread of cooperative evolution winds its way through our narrative. A look at its title might suggest to a casual browser that we are joining the authors of the *How Darwin Got It Wrong* genre. Most definitely we are not, although we acknowledge that, occasionally, Darwin fell into error – for example, thinking that evolutionary change is always slow; attempting to explain biological inheritance without knowledge of genes; and limited recognition of the effects of inevitable changes in the environment.

Darwin's genius was to recognise the underlying principle of *natural selection*: uncontrolled reproduction produces more offspring than the environment can bear, so they strive to find ways to a share of scarce resources. Offspring do not exactly resemble their parents. They vary, and this variation is the raw material of evolution. We perceive, by hindsight, those that have turned out to have had, statistically, a better chance of survival are more likely to be those that live long enough to contribute to the next generation. Then the process of selection is repeated. Natural selection is an iterative process; many trillions of repetitions have taken life from the first cell to the blue whale and the human being.

Those who are interested in Darwin's theory of evolution often wish vainly that they could, in the modern idiom, 'channel' the great man himself. Few of us could take this idea further. In 2000, however, in *Dear Mr Darwin*, the late Gabriel Dover started an imaginary correspondence with Darwin – from his last resting place in Westminster Abbey. Darwin, obliging as ever, 'commented' on the discoveries of the last 150 years.

Like Dover, we have tried to imagine what Darwin would have made of the last 150 years of biological discovery. We do this with due homage to Darwin's original vision of a collaborative world, a vision that had become obscured, even in his own time. This book attempts to explain how Darwinian evolution by natural selection works, and where it might go next. It tells stories of cooperation between different organisms with different genetic backgrounds and between organisms that share the same genetic heritage. It tells of the origin of modern cells, of multicellular organisms, of symbiosis and of social collaboration. Along the way it explores the crucial influence of human interpretations of natural systems, and of solar energy in promoting organisation in the systems through which it flows. Finally, it deals with our own origins and the origins of our social choices.

Evolution thus began when life's first, ancient, cells spread across the globe with all the time in the Solar System at their disposal, probing the Earth, seeking the conditions for their own survival, leaving behind the trail of their explorations inscribed in their DNA – some copies of which you, the reader, still have in your cells. The appearance of the modern cell is the result of the first great act of cooperative evolution between ancient cells. It is the first spinning of the evolutionary threads that are now woven into a planetary phenomenon.

With time, some of these threads intertwine. Complexity increases. Sometimes, however, some seem to stop – removed from an endlessly changing environment to remote ones of great stability. These become the living fossils: the coelacanth, which was common 400 million years ago, survives in the deepest oceans; the recently discovered Wollemi pine is a 200-million-year relict from the old continent of Gondwana. Others turn back on themselves, like the white, blind denizens trapped in the ocean depths or in dark, underground caves. Yet others, disadvantaged by the changing conditions to which it is beyond their capacity to respond, die out. The survivors spread outward; they meet each other and, invigorated by the increased opportunities for cooperation so afforded, diversify and spread further. They encounter obstacles and grow around them, gradually occupying all available livable space, from the top of Everest to the bottom of the Mariana Trench, kilometres above and below the surface of the continents.

Some threads travel through time together, join with other threads to make their own domain of life forms. Some of these are familiar. Animals with backbones or those, like insects, with their skeletons on the outside; plants with cones; plants with flowers; plants that still live in water. Some, like the lichens, or the single-celled Protista, or the surviving colonies of ancient cells, are everywhere and you may never even have noticed them. Eventually, they form a continuous, 3-billion-year-old net of interaction covering the globe. It is everywhere and touches everything. It is the tangled bank writ large.

All organisms that ever lived spend much of their time acquiring sufficient food and other resources and establishing productive relationships with others in their environments. Many find themselves in the right place at the right time, with the right portfolio of attributes, but others are less fortunate. They find themselves in the wrong place, where the appropriate conditions just do not come together in the right way.

Sometimes, although they seem at first to be lucky, they do not form the relationships with others that are necessary for prolonged survival. Frequently, however, an association between two dissimilar kinds becomes a permanent partnership that enhances the chance of their mutual survival; as the algae and fungi that form lichens, or the lions and hyaenas that share the kill on the African veldt – or the interdependent urban and rural modern humans.

It may be that breeding regimes are such that they leave fewer offspring than some of their better-resourced neighbours. Sometimes they survive with help from others of their own kind, like shoals of fish or flocks of birds or prides of lions. In the case of human communities, 'it takes a village to raise a child'. This is our argument: all the great evolutionary successes depended, primarily, on cooperation rather than competition.

The book moves on to explore how Darwinian evolution by natural selection works, and where it might go next. It tells stories of cooperation between different organisms with different genetic backgrounds and between organisms that share the same genetic heritage. It tells of the origin of modern cells, of multicellular organisms and of symbiosis. Along the way it explores the crucial role of solar energy in promoting organisation in the systems through which it flows. Finally, it deals with our own origins and the origins of our society.

The impact of social Darwinism on humanity in all its aspects is beyond the scope of this story, which is about one aspect of the evolutionary process. There is a plethora of books that are concerned with the way the human world changed after the publication of *Origin*. For those interested, we recommend Desmond (1989), Desmond and Moore (1991) and Browne (2003). The website of the Darwin Project is at www.thedarwinproject.com.

We tell the story of cooperative evolution with due homage to Darwin's original vision of a collaborative world that has become obscured. We are so very fortunate to have access to another 150 years' worth of biological research that was not available to Darwin. We have tried, as Gabriel Dover did in his *Dear Mr Darwin*, to imagine what Darwin might have made of it.

2

ALL KNOWLEDGE
IS METAPHOR

In which we argue that we can know nothing 'for certain'. The best thinkers generate the best metaphors. Evolution is a particularly powerful one.

As they begin their careers, budding biologists are, quite rightly, cautioned against the mortal sin of *teleology*. This word is derived from the Greek *telos* meaning 'end' and *logos* 'reason'. Its extreme form can be expressed by an example, thus: 'the purpose of evolution is to create humans'. Few will deny that the process of evolution has produced humans, but the process itself is not concerned with producing humans. There is no reason to suppose that if the whole of evolution were reset at the beginning and rerun, we would get exactly the same outcomes. We might see similarities, because convergent evolution is a well-studied phenomenon, but not the same.

Yet evolution *appears* to be purposive. If you study it, you get a feeling of direction and organisation increasing with time. These are called trends – trends towards multicellular organisms, towards increasing size, towards increasing intelligence and so on. The trends illustrate ways of succeeding in the universal objective of staying alive as long as possible, to improve chances of producing viable offspring that will continue the process.

Consider this sentence:

> The view through the window shows the leaves that are scattering as the wind blows them across the road and the small clouds that are racing across the sky.

But wait a minute. Notwithstanding the conventional images of mythical winged beings in the corners of ancient maps, winds do not 'blow'. Only humans do that. Neither do clouds 'race'. Racing has implications of competition against others, or against time. Oh, and time does not race either, the clock 'tells' you that it ticks along at steady rate. But a clock 'tells' you nothing … And so on. Human language is full of conventional metaphors that are clearly understood by those who speak the same language.

From its beginnings in the seventeenth century, scientists have worried about the meanings of words and their ambiguity, particularly when naming plants or animals. The 'father of taxonomy', Carl Linnaeus, did the scientific world a great service by introducing the system of binomial classification. For example, before Linnaeus, English people might talk about the blackbird (or, in the next county, the ouzel) and the song thrush (or mavis, across the border in Scotland), without being sure that they were referring to the same bird. After Linnaeus gave them the rather unfortunate generic name of *Turdus* coupled with different specific names, they became, respectively, *T. merula* and *T. philomelos*. No more ambiguity and, thanks to Linnaeus's system of classification, we can be sure we are all singing from the same song sheet when we praise the calls of these two common songbirds.

Scientists like their own 'language' to be unambiguous. For example, 'acid' is a class of molecule with very specific properties and they do not really like it if people 'drop acid' or make 'acid' remarks or write with a pen 'dipped in vitriol'. Vitriol is the old name for sulphuric acid. Perhaps ambiguity might be removed by writing its formula instead: 'She writes with a pen dipped in H_2SO_4'.

It doesn't work – chemistry kills prose!

The ideal of science is to be objective, to describe the world as it 'really' is. This is impossible when writing in any natural language. The everyday words that we use carry too much baggage for objectivity. What is a rose? Gertrude Stein wrote that 'a rose is a rose is a rose'. But is it? Because it is also a member of the family Rosaceae, a perfume, a beautiful young woman, a symbol of true love, a device attached to a watering can, a compass and the badge of the medieval House of Lancaster. Change its colour from red to white and it is the badge of the medieval House of York. *The Name of the Rose* is a book written by Umberto Eco. We determine what is meant by 'rose' by its context.

Language is therefore an imperfect method of communication. In fact, Darwin used it to his advantage when discussing his ideas with non-scientists. In *On the Origin of Species* he refers to 'a Creator', rather than 'God', at the end of the famous last 'tangled bank' paragraph. In his letters, he subsequently regretted even this.

We have already, in this chapter, discussed the 'biological sin' of claiming that the end point of a process of biological adaptation is the cause of it taking place. Two other biological sins can be joined with this one: they are anthropocentrism and anthropomorphism. Anthropocentrism is the conviction, held by many religious groups, that humans are the focal point of the universe. The universe was constructed for us, and our planetary system is designed for our convenience. Genesis (1:26) even tells us that humans have:

> dominion over the fish of the sea, and over the fowl of the air, and over the cattle, and over all the earth, and over every creeping thing that creepeth upon the earth.

One has only to look through a microscope or a large telescope to doubt this proposition.

Anthropomorphism is imagining that inanimate objects around us are alive and that animals are really little (or large) human beings. Our ancestors saw faces in the fire, animals in the clouds, gods in the trees and stones, and were fearful. Our children are particularly prone to anthropomorphism and live vicarious lives through their teddy bears and dolls, honing their developing empathic skills. The familiar concept of 'Mother Nature' herself is an anthropomorphism. She is a fictional demigoddess who is somehow loving, nurturing and compassionate. This is a view of life far from the true state of affairs.

We were told, as students, that if we felt an urge coming on to attribute purpose to evolution, or if we felt that our ancient inclination to anthropocentrism and anthropomorphism was emerging, we should, for our own biological respectability, preface it with the words *it is as if …*

There are a lot of places in this book where purists would have us write *It is as if ...* If we did so, however, it would be repetitive and tedious for the readers. Instead, we crave indulgence and ask them to take the qualification for granted where we have omitted it. Evolution does not have 'purpose' in the human sense but it is convenient to write as if it did. Living organisms do not have human purposes and human intelligences, but it is convenient to write as if they did. Later on in this chapter we refer to James Lovelock's metaphor of Gaia, only four letters, but used to describe 'the self-organising, self-maintaining, cooperative, global biosphere that interacts with the physical and chemical world'. Four letters doing the work of 110 that are themselves shorthand for the greatest overarching entity on Earth. How convenient is that?

All science is metaphor. Charles Darwin's metaphor of natural selection as the origin of species was a globally momentous one. Those two words encapsulate the essence of the evolutionary process, meaning 'it is as if Nature were selecting from among the offspring of each organism'. From which it follows that 'it is as if Nature were capable of choosing' and 'it is as if Nature were an individual, the sum total of biological activity'. Or, to quote again from the 'tangled bank' paragraph in the *Origin*:

> these elaborately constructed forms, so different from each other, and dependent upon each other in so complex a manner, have all been produced by laws acting around us.

'Dependent upon each other in so complex a manner' paves the way for two further great metaphors, this time of the twentieth century, one of which has great relevance for us in the twenty-first. One explains the evolutionary origin of the modern, non-bacterial cell, the other the evolutionary origin of the planet itself. We will revisit them in Chapter 3.

We have already discussed the role of metaphor in language. Now we go further: all reality is metaphor. The metaphor of natural selection facilitates our description of the properties of evolution. Two important ones are that it self-organises and encourages cooperation among the life forms it shapes. Given these characteristics, we conclude that it is as if the golden thread of Darwin's cooperative evolution has woven the biosphere into a network that enfolds the planet to create James Lovelock's Gaia, another metaphor that commands our respect.

If you doubt the power of reality as metaphor, consider particle physics. The concept of the atom itself is a metaphor, one that has been refined over millennia. No-one has ever seen a single atom. Democritus (c. 460–370 BC) is first recorded as having imagined some of its properties; obviously never having seen one he could only say 'it is as if' the atom were the smallest indivisible bit of matter. Pure metaphor. Others refined it and, at the beginning of the nineteenth century, Dalton distilled the idea of an atom into a set of axioms that still hold true today.

Subsequent atomic models ('model' is another name for a metaphor) have represented the atom as a plum pudding with the electrons as the plums or, by analogy with the Solar System, a 'planetary' system in which electrons orbit a nucleus made up of protons and neutrons. That last sentence is dripping with metaphor. The 'planetary system' appears, much modified, in most of today's textbooks. We can continue this metaphor hunt, as physicists do, into particle physics with 'quantum', 'string' and 'knot' hypotheses, or argue that subatomic particles have the qualities of 'charm' and 'spin' and 'up' and 'down'. All metaphors. So is the idea of any other organisational unit from a cell to a city. Indeed, the city is often used as a metaphor to explain the functions of a living cell. It works well both ways.

Metaphors of evolutionary change are many and varied. The 'missing link' in the 'great chain of being', the 'survival of the fittest', 'adaptive radiation', 'the apple does not fall far from the tree' and even DNA: these are metaphors, all of which apply to evolutionary change, that have entered our language until we have forgotten that they are indeed metaphors.

Take, for example, an enzyme. An enzyme is a component of the cell that cannot be detected *unless it is doing its job* of enabling a chemical reaction. Enzymes are many and have been an essential part of life from the beginning – and also a part of whatever went before fully living systems. Indeed, it is enzymes that enable life to control one of the essential resources of life, energy flow, discussed in Chapter 7.

All life depends on cells. This includes viruses that may be precursors to or derived from cells but need to enter a cell to multiply. Unlike viruses, cells are self-contained entities that are capable of carrying out all the functions of life. As we will see, there are fundamentally two types of cell. There are ancient cells – bacteria and blue-green algae – and there are modern cells, the ones that are found in all other living things.

In the nineteenth century, German physiologist Wilhelm Kühne (1837–1900) discovered that non-living extracts of cells could carry out chemical reactions. He called the unknown agents 'enzymes' and so began the science of biochemistry. Thousands of enzymes are now listed by the International Union of Biochemistry and Molecular Biology's Enzyme Commission.

An enzyme is a molecule that promotes a chemical reaction while remaining itself unchanged. They have many practical applications. You may have encountered them in detergents for removing 'stubborn stains'. They are also used in textile and paper manufacture and in foods and beverages. Rennet is an enzyme from calves' stomachs. It was used traditionally for making milk junkets and can still be bought at the supermarket.

Victorian biochemists thought that enzymes were the very stuff of life and, indeed, they were not far wrong. All cellular activities are accomplished by means of enzymes. In the twenty-first century we know that the complex protein molecules that form the majority of biological enzymes have themselves evolved from simpler molecules. But has anybody ever *seen* an enzyme? They have seen solutions that contain them. They may have seen their shadows in electron micrographs. They may have seen crystals of protein enzymes, but sadly, such crystals have no enzymatic properties – these do not appear until the crystals are dissolved in water with an appropriate mixture of other molecules. The concept 'enzyme' is a *quality* displayed by a molecule only under the right circumstances; and is an extremely useful metaphor.

An elaborate set of metaphors links the ideas associated with a modern cell with the way that the first cities came about. This is because the biological imperatives for both cell and city are the same: the control and coordination of the basic necessities of life. The city, of course, as a human construction, has social imperatives as well.

Historian of ideas Lewis Mumford developed a series of criteria that helped explain the complexity of a modern city. First, a city needs a granary, to hold the community's corn safe and dry so as provide seed and food for the following year. At its earliest, the seed store was just a deep hole in dry ground. Then, because of the danger of flooding, communities began to make large clay pots in which to store the seed above ground. Later still came barns and silos. When money came into use (a metaphor for seed corn and future prosperity) the 'granary' became a savings bank.

Second, there was the need for a holy place to provide a community focus and to encourage social cohesion. There, rituals were performed to appease a pantheon of gods who were considered responsible for environmental change, life and death. The holy places began to acquire structure, wood and stone circles. Today, there are great churches, sacred spaces designed to preserve the valued objects of the people. Associated with them were, first, the great libraries and, later, museums.

Third was the control centre of the budding city, a chieftain's hall, a queen's royal palace or a war lord's citadel, where the decisions were made about the obligations of citizens and means of defence for the community. It was a place where the rulers and the community could meet and agree on future plans: a centre of government, a parliament. The power of the community was vested in the priests of the temple and the guardians of the citadel, those who controlled the food supply and other resources. Strong people were required to wrestle with gods, to defend the city and husband resources. Communities began to organise themselves into a class system, with serfs at the bottom layer and the priests and royalty at the top.

As we now know, the nucleus of the modern cell coordinates cellular activities so that they are responsive to environmental change (appeasing the gods!) and defence. These are analogous to the roles of the church, the citadel and the library. There are membranous structures, the mitochondria, that are the 'powerhouses' of the modern cell. The cell membrane (the city wall) is continually taking in, through its gates, nutrient molecules to be broken down by metabolic enzymes (serfs, who do all the hard work). A transport system delivers the molecules to the powerhouse, to act as fuel from which to derive the energy that is used to drive all cellular functions.

The other major type of modern cell, the plant cell, taps into the primary power source, the sun, by means of chloroplasts. These are collectors of solar energy. Plant cells do not need a continual source of proteins and carbohydrates from outside. They can make their own as long as they have access to carbon dioxide and a source of inorganic nitrogen (this is why ammonium nitrate is such a good fertiliser – it is 35 per cent by weight nitrogen).

Thus, cells and cities can be made to fall within the same set of metaphors. For modern cells, the best comparison is with a medieval city at the beginning of the Industrial Revolution rather than the giant conurbations of today. It is better to think of those as clusters of cities, or 'multi-urban' by analogy with 'multi-cellular'. Almost all the functions of a pre-industrial city are represented in the cells that make up an individual life form. Or is it the other way around? Are the functions of the cell represented in the city?

The city and the cell at maturity are steady state systems, with the imports ideally balancing the exports, and for the city, immigration balancing emigration. When the balance is lost, then cities become disorganised and cells grow old and die. One can go on making these comparisons. And the reason for this is that all forms of life organise themselves around their need for resources. At the most basic level, the needs of people are the needs of cells.

Now here are two metaphors, of a type more generally known as fables. Stay with them, for they are part of the argument! They demonstrate that biological and social phenomena, when viewed at vastly different levels of magnification, can show remarkable similarities.

Fable 1. Ken and the Grex

At Harvard University, a young man, Ken, was trying to define a research question about the behaviour of a type of single-celled amoeba. He picked up a Petri dish filled with nutrient agar gel, suitable nourishment for the little organisms. He then reached for a culture tube containing amoebae that he'd collected and added a drop to the agar plate. It spread out over the surface and he settled down with a microscope to take observations over the next two or three days. Under the microscope, the cells of the amoebae appeared as small silvery blobs moving slowly around on the gel. He noted that the blobs were widely distributed, sometimes as individuals, sometimes in small clumps.

All at once, he realised that all the individual amoebae in his field of view had been slowly moving – except that one had stopped and now appeared to be attracting or signalling, somehow, to the others. After a while he noted that more and more blobs seemed to be forming small associations or clumps. Gradually, all the amoebae coalesced into streams and headed towards the amoeba that had first started signalling. Just like, the young man thought whimsically, a huge crowd of people, viewed from the air, heading into a major baseball game. What was the attraction? A series of clever experiments showed him that the one stationary amoeba was emitting a powerful chemical attractant. After a while, a large clump formed. Eventually the clump became an independent slug-like creature. It started moving in a coordinated way in one direction and he recognised it as a slime mould. The student wrote an important thesis on the self-organising capacities of the 'grex' as he named the strange slug-like accumulation of amoebae.

Fable 2. The Alien and the Earthlings

A young Alien from a not too distant planet, having taken a self-teaching unit called Earth studies, was trying to learn more about the planet Earth and to define a research question. Her space vehicle was parked in stationary orbit over a large collection of buildings that the Earthlings called a city. In the centre was a huge circular enclosure that was, at that moment, empty of Earthlings. The Alien settled down with her electron telescope to take observations of the round space that she had mentally called an arena. Over the next two or three hours, the Earthlings appeared as small coloured blobs moving all over the city. The Alien noted that the blobs were widely distributed, sometimes as individuals, sometimes in small clumps.

Then she noticed that there was a general movement towards the arena. Obviously, there was some kind of attractant to bring the blobs together. At first, only a few Earthlings were slowly moving around the enclosure. After a while she noted more and more Earthlings appearing. Suddenly she realised that *all* the Earthlings in her field of view were slowly moving to what appeared to be gaps in the enclosure. They then became the focus of attention of all the others. They appeared to be signalling, calling to each other. Gradually, all the Earthlings coalesced into streams heading through the gaps.

What was the attraction? She noted that, after a while, the large gathering had appeared to coalesce into a single clump-like group of individuals around the rim of the arena. Later still, the rim had turned into a single creature, responding as a whole with coordinated sounds, and wave-like motions. The waves seemed to correspond with the actions of a small group within the centre of the enclosure or arena.

The Alien went home and wrote interesting stories about the self-organising capacities of the Earthlings.

At the end of the twenty-first century, the Earthlings discovered a tiny, inert satellite, about 5 metres in diameter, in stationary orbit around Earth. When its recordings were deciphered, it was found that the ant-sized Aliens' interstellar achievements derived from their ability to form a conscious collective intelligence.

The Alien and the Earthlings is an imagined story, so it is strictly science fiction. *Ken and the Grex,* though, is science 'faction', almost true. There *was* such a PhD student in 1933; his name was Kenneth Raper, and he discovered the interesting little slime mould, *Dictyostelium discoideum* (Raper 1935). He died 54 years later, a Fellow of the American Academy of Sciences, replete with scientific honours. His slime mould has become a model organism (a metaphor) for studies of the evolution of self-organisation, multicellularity and embryonic development.

And the moral of our two fables, as the duchess might have said to Alice when wandering through Wonderland, is that *Dictyostelium* and *Homo sapiens* both start off as individuals and organise themselves into complex structures. Cells and people respond in similar ways – but a person has another layer of history because, right at the beginning, they start off as a single cell, a fertilised egg, and self-organise into a human being and later, as a human being, become part of a self-organising social system.

Further, cooperative evolution has a habit of repeating itself. What worked once can work again – at different levels of magnification and complexity. The modern cell is a cooperative construct of three or four ancient, bacteria-like cells working in partnership (see Chapter 3). It was a great evolutionary success. If it worked with ancient cells, why not try modern cells? And it worked again. We find modern cells cooperating to make all the plants, the fungi and the animals. Take another step, and we have organisms of one species – bees or humans – working together in colonies and developing social systems. If that works, why not try collaboration between organisms of different species? This is cooperative evolution, an intrinsic property of life on Earth and the subject of this book.

To return to the fables, the view through the microscope is like the view the imaginary alien had of the Earth through her telescope. Eventually all the earthlings, like amoebae, coalesce to form a 'grex' or crowd, which grows and grows as more join in. Now they can enjoy a much higher level of connectivity – that is to say, information transfer – and go to sporting events, interacting with each other and their environment to produce a Mexican wave.

Dictyostelium slugs and towns have this in common. In the beginning, there is no central slug-organising cell or group of cells. The slug grows until it reaches an optimum size and it starts to move. Arriving at a suitable place, the slug stops and differentiates further as cells assume different functions. And it also appears that *Dictyostelium* performs a sort of bacterial farming, because it transpires that these cells have brought bacterial spores with them to seed their new 'pastures'. Its amoebic citizens in spore form undergo a diaspora (!) and with their precious cargo of bacterial spores blow away to found new communities.

The city starts as a settlement of a few people, then becomes a village and self-organises itself around some basic human needs, grows into a town and then a city that persists for a time until it reaches the end of its cycle and disperses.

To carry biological analogy even further, individual humans, during social development, pass from a single-celled stage, through a multi-celled stage that is analogous to a village, represented by the very early embryo. The 'village' rapidly becomes a small 'town' as it passes through stages of tissue and organ formation, and finally becomes a human being.

The evolution of human societies begins with the cellular family, which enlarges with simple cooperation between family members followed by recruitment of outsiders, to the point of permanent differentiation, where individuals began to specialise in specific tasks.

Towns, when they grow naturally, rather than being laid out by surveyors, develop similar characteristics to one another and are usually sited to take advantage of some important local resource, such as water and minerals. In early human communities, gold was not so important but flint, copper, tin and iron soon became so. The flint mines of Grimes Graves, in Norfolk, England, are evidence that flint had a considerable pulling power. The flint mines were worked for about 300 years, more than 4,000 years ago. Grimes Graves became an industrial centre that exported worked flints to Europe. Still in England, copper and tin mining in the early Bronze Age also created important centres.

When the smelting of bronze started, what we might call the 'tissue' stage arrived as, within their communities, people of like purpose began to associate. Populations stratified into classes, some charged with defence of the commons, who worked closely with a priestly class that interceded on behalf of the people for the favour of the gods. Others plied the various trades that made up a Bronze Age community, and, in Homer's words, 'By mutual confidence and mutual aid great deeds are done, and great discoveries made'.

Communities continued to increase in size and became tribes and city-states. Still clinging to the biological analogy, this is the time of organ formation. As a biological organ is made up of cooperating tissues (made up, in turn, of cooperating cells) so different trades took on different roles. The leather industry, say, diversified into tanners, cordwainers and cobblers, harness makers, jerkin makers and so on. Groups of artisans working in a common endeavour formed into societies and organisations to protect their knowledge and livelihood. Out of this came the medieval guilds.

Figure 1 summarises some of the unintended consequences of being human. As connectivity between individuals increases, so do their institutions increase in complexity. Thus, guilds grew out of the fraternity of, say, stoneworkers, and exemplify the way that barriers were gradually being erected between classes of people.

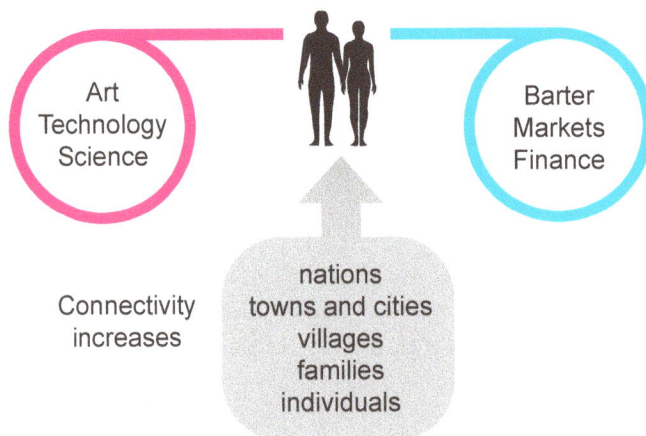

Figure 1. The unintended consequences of being human.

They became formalised and the barriers were strongly enforced by the guilds who protected their knowledge obsessively and even secretly. They erected professional hurdles that acolytes had to climb over before they could be admitted to membership. In the modern world they still exist under different names – a university is such a one, although not secret – and are essential to the smooth running of nations.

Returning to the two fables, they show that views of life at different levels of magnification offer the same general picture. Like that classical example of a fractal, the coastline that offers the same general characteristics at every magnification, biological systems also offer the same general view wherever you look. Rainforests on different continents are created from different components, but a rainforest is recognisably a rainforest, whether you are in Australia, Africa, the Americas, Asia or Europe. Rainforests have the same general aspect: they smell the same, they feel the same and a tree is recognisably a tree, wherever you are. Rainforests all 'work' in the same way. The differences lie in plants and animals that each occupy similar ecological niches. In your mind's eye, follow this thread of biological cooperation and travel back through layers of time, stopping every so often to look around. What you see at every stop may be a dramatically different stage of forest development.

It is no accident that human behaviour maps onto biological systems. First, people are made of the stuff of life and this chapter is intended to illustrate and emphasise the repetitiveness of living systems. What worked once can work again. It also emphasises the layers of evolution. In the beginning, the world was solely inhabited by ancient types of cells that were similar in form to modern bacteria. Three or four of them got together to make the modern cell, the sort of cell of which we are made. Modern cells got together and made animals and plants. The whole of organic evolution was implicit in that first cooperative assemblage of ancient cells. Third, it illustrates the important phenomenon of self-organisation. Cities and slime moulds are self-organised; so are ant, wasp, bee and termite societies, so are towns and cities, so are ecological systems.

All of the animal phyla that we recognise today existed half a billion years ago, in the Cambrian period. At that time, the only photosynthetic plants were the blue-green algae, and seaweeds. All other plant groups were dependent on land colonisation, for which there is evidence, 450 million years old, in the form of fossilised plant spores. Every organism since that time has been brought about by evolution playing with what has gone before – mixing, stretching, squeezing, twisting, duplicating, aggregating, slowing or speeding development, or sculpting with programmed cell death. Living things are like 'the Colonel's Lady and Judy O'Grady', sisters under the skin, as Rudyard Kipling remarked in another context. But they are far more than that. They are all obeying the imperatives of life set down with the birth of the first cell.

In this chapter, we have briefly explored the power of metaphor in science and in human affairs and have introduced the stories that will emerge as we follow evolution from Charles Darwin's earliest ideas to Lovelock's formulation of the Gaia hypothesis. Earth's history is a series of tales of increasing connection and collaboration. These concepts, of course, are metaphors themselves but are the best that philosophers can do, given the state of knowledge. Reality itself is a metaphor and we have to explore it through the eyes of scientists, historians and sociologists.

3

INTELLIGENT EVOLUTION AND INTELLIGENCE

In which we observe that the evolutionary journey from ancient cell to humanity is marked by increasingly intelligent life forms. The human brain is the most highly interconnected, and therefore internally collaborative, entity on the planet.

In higher animals, consciousness has evolved and can be brought to bear on the problem of survival. With consciousness, true purpose emerges. Humans can set out to make things, having clearly visualised the end products. It seems likely that this ability to foresee the outcomes of certain activities is shared by some primate, bird and octopus species. There may be others, but more than 99 per cent of living creatures are plants and do not have this form of consciousness – yet they are able to solve the problems of staying alive.

Darwin was the undoubted biological colossus, the nonpareil, of the nineteenth century. The twentieth century, however, held many giants, two of whom concern us here. One was Lynn Margulis (1970, 1991), whose work on the symbiotic origin of modern cells from ancient cells forms one of the starting points of this book. She wrote, with co-author Dorion Sagan (1997), that:

> The view of evolution as a chronic bloody competition among individuals and species, a popular distortion of Darwin's notion of 'survival of the fittest,' dissolves before a new view of continual cooperation, strong interaction, and mutual dependence among

life forms. Life did not take over the globe by combat, but by networking. Life forms multiplied and complexified by co-opting others, not just by killing them.

The other is James Lovelock (2013) with whom Margulis was an early collaborator. Lovelock is remarkable both for his longevity (he is still working at 100, as of the moment of writing) and the fact that he forged a career as an independent scientist. Even more remarkable is his world view.

Both Charles Darwin and James Lovelock created powerful metaphors to express and establish their ideas. Darwin took a concept that was almost unthinkable – that all life forms are shaped by evolutionary change – and clothed it with a plethora of scientific observation. His formulation of the metaphor of natural selection changed the world.

James Lovelock took Darwin's metaphor, and realised that the Earth was indeed a living world: a complex, self-regulating and self-organising entity.

Lovelock's personal epiphany occurred when he saw the first photographs of Earth from space. He saw a planet that was blue because of the oxygen in the atmosphere. It should not be there – on thermodynamic grounds alone, it should be tied up in nitrates and carbon dioxide. The awe-inspiring conclusion is that life itself regulates the composition of the atmosphere. Oxygen is there *only* because of the oxygen-producing activity of all the green plant cells on the planet, the photosynthesisers. Lovelock explains that if the laws of chemistry are all that matter, all the oxygen and most of the nitrogen in the atmosphere ought to have ended up in the sea, combined as nitrate ion.

> The air we breathe cannot be a very fortunate one-off emanation from the rocks; it can only be an artefact maintained in a steady state far from chemical equilibrium by the biological process.

It was not always so. The first world of the ancient cells was anaerobic, oxygen-free. Then one type of ancient cell learned very early to make use of the carbon dioxide that was already in the atmosphere. It became a source of nutrition, used to make simple sugars. Oxygen was produced as a by-product of this biochemical process of photosynthesis, and it was poured out into an otherwise oxygen-free atmosphere for many millions of years. As fast as the cells produced oxygen it was mopped up by geological strata containing iron, causing them to 'rust' and form the ancient 'red beds' that are now a source of iron ore. Only when all the available iron had finished 'rusting' could the oxygen begin to appear in the atmosphere. As life continued to evolve, the amount of oxygen in the atmosphere increased and, with many fluctuations, finally reached today's level. From the point of view of all the other cells that were not photosynthesisers, it was the first great act of planetary pollution, to which they were forced to adapt. One way to adapt was to cooperate with the photosynthesisers.

The richness of metaphor is displayed when creative writers link the conceptual to the physical world. Lovelock happened to be friendly with the writer, William Golding. One day, he explained his idea to Golding and wondered if he could suggest a name. Golding suggested 'Gaia' for Lovelock's vision of the world. Gaia was an ancient Greek goddess, a personification of the Earth and one of the deities at the beginning of all things.

Lovelock had realised that the biosphere, the sum total of life on Earth, continually interacts with the atmosphere on the one hand and the geosphere on the other. Within the biosphere, numerous interactions occur that help to stabilise the whole system. This living system – we may now call it Gaia – is a vast, but not limitless, self-regulating entity that humanity is on the verge of overexploiting and polluting, putting its own continued existence in peril.

Not everyone recognised the importance of this discovery; indeed, Lovelock was accused by many of trying to resurrect a caring 'Mother Nature'. This scepticism explains why he felt more at home with those scientists – physicists, chemists, biochemists and engineers – who understood the centrality of the concepts of cycles, of feed-forward and feedback, up- and down-regulation, that are the marks of energy flows. He later remarked that his references to Gaia as a living organism were intended as shorthand, a metaphor, not to be taken literally, to describe an entity of globally interactive feedback systems displaying the capacity for self-regulation or homeostasis. This is illustrated in Figure 2, in which humans are represented as part of the biosphere. It was when they became too deeply involved in the biosphere that trouble started; the Industrial Revolution threw a monkey wrench in the works and disturbed the steady state.

Lovelock stated, rather irritably:

> Of course the Earth is not the phenotype of any species of organism, but the coupling between all the individuals of the planet and their material environment (that) results in a homeostasis similar to that of the nest.

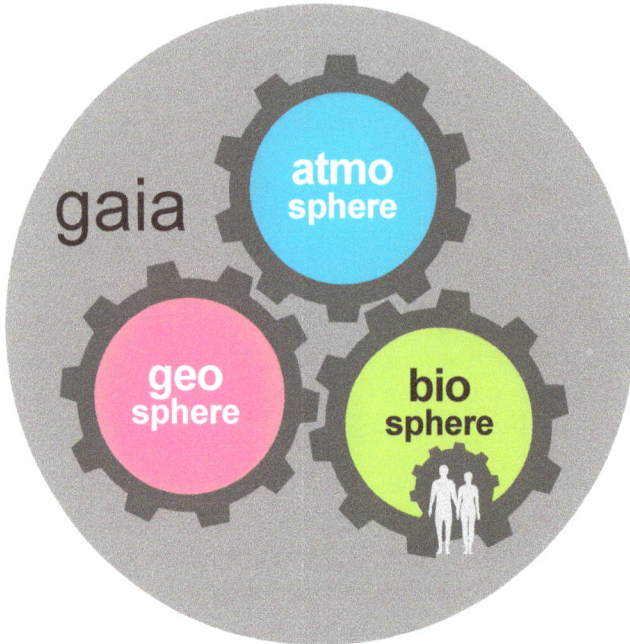

Figure 2. Humans in the biosphere.

'The nest' is a reference to the termite mounds, ants' nests and other structures of social insects that allow them to maintain a steady internal environment (homeostasis). Illustrated in Figure 3 are two nests of cathedral termites: the unpredictable emergent consequence of the self-regulation of termites going about their daily business.

Newton's third law of motion states that 'for every action there is an equal and opposite reaction'. Le Chatelier rephrased it for chemistry and biochemistry as a principle, in the following terms: 'if a system that is in chemical equilibrium (long-term balance) is subjected to a disturbance it tends to change in a way that opposes this disturbance'.

The carbon dioxide concentration in the atmosphere is at present increasing, and so are average global temperatures. As a result, weather is becoming more extreme and unpredictable; the poles, continental snow caps and glaciers are melting; the seas are becoming more acidic; ocean currents are being disrupted. It is difficult not to conclude that we are observing the global steady state adapting to new circumstances, and that outcome will be to establish a new steady state for Gaia. Humanity is contributing to those circumstances, so Gaia's response carries a threat to humanity, for the new steady state may not be one friendly to humanity's continued existence.

**Figure 3. Termite mounds: the unexpected consequence of termites'
self-regulation.**

Source: Photograph by Chris Bryant.

It is important to understand the difference between the terms 'steady
state' and 'equilibrium' when used by scientists. The steady state is an
'open' system, such as the river illustrated in Figure 4a.

1. An open system
in **Steady State**

2. An enclosed system
in **Equilibrium**

Figure 4a. Equilibrium and steady state systems.

A river flows along its course continuously. If you stand in the same place on a river bank to view it on consecutive days, the scene will look very much the same. During that time, however, the part of the river that you see will have been replenished many times by the water passing through. As Heraclitus is credited with saying, about 500 years BC, 'you cannot step twice in the same river', because the bit of water you stepped in a few seconds ago is now several metres downstream. If, however, you enclosed some of the water in a large can, it would not move. An enclosed system is a special case of the steady state called equilibrium. Nothing happens at equilibrium.

These two conditions are again represented in diagrammatic form in Figure 4b. On the left, *a* and *b* are two rivers that join to make a single large river *ab*. The lesser rivers, *a* and *b*, are draining water from a range of hills into the larger river, *ab*, made from their confluence. River *ab* continues to the sea where it discharges its water. As long as it rains in the hills, the rivers will flow – a steady state. If there is a drought, eventually river *ab* will not make it all the way to sea and the system begins to stagnate. There is no movement of water.

Now let us change the metaphor using the same figure. We are now looking at a series of reactions within a living cell; *a* is a chemical produced by one set of reactions, *b* by another. It is an open system within the cell and they can be replenished. The two can react together to produce *ab* which, in turn, can react with a variety of other substances to produce a range of products. The open system is full of possibilities for maintaining a steady state and even growth.

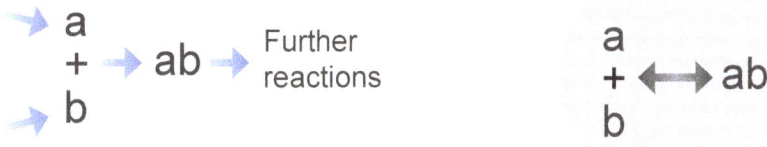

Figure 4b. Open and closed systems.

Compare this with the closed system inside the can. The same chemicals, *a* and *b*, are there, but as the can is closed there is no hope of getting any more. They can react to form *ab* but there is nothing for *ab* to react with. Its only option is to collapse back into *a* plus *b*, which is a futile equilibrium. Nothing interesting can happen until the can is opened and ancient cells can set to work on its contents.

The mechanism that builds changes into the biosphere is Darwinian natural selection. Natural selection is a testament to the power of iteration, which allows changes to build on themselves. Some examples of patterns generated by iteration are shown in Figure 5. Classic fractal patterns are generated by repetition at different levels of magnification. When the broken outline of the edge of an island, say, is viewed under greater and greater magnification, at each level, bays, coves and inlets are progressively revealed as repetitions of the larger pattern.

Figure 5. Fractal images.

Simple, repeated iteration in such patterns, as in the human lung or a branching tree, creates structures of high complexity. A fractal pattern thus stands as a simple metaphor for the structural complexity and repetition of evolutionary change. Darwin's evolutionary theory, of continual change by natural selection among life forms, is an iterated process in time – one that has been going on for 4 billion years.

The effectiveness of natural selection derives from the fact that organisms do not exactly resemble their parents. Some of the variants prove to be less able than others to survive in the environment in which they find themselves. They do not necessarily drop dead – they simply do not have resources enough to out-breed the lucky ones, so their populations dwindle. Their ultimate extinction may take many generations, and efforts to bring back species that have reached the ends of their lines, with population sizes that are no longer viable, may be doomed to failure. Dwindling in this way was probably the fate of the Neanderthals, who, though intelligent humans, were unlucky enough to coexist with more modern humans and now live on only as part of our DNA.

'Intelligent' is a wispy, smoky concept, and yet we all think we know what it is. Go to a dictionary and you find all sorts of circular definitions. One is 'having a high degree of intelligence and mental capacity'. Legg and Hutter open their 2007 paper on defining intelligence with the words:

> Despite a long history of research and debate there is still no standard definition of intelligence. This has led some to believe that intelligence may be approximately described, but cannot be fully defined.

They go on to examine no fewer than 71 attempts at a definition of intelligence.

In this book we adopt the view that an intelligent action is one that leads to a good outcome for the entity that undertook it, whether it has consciousness or not. Thus, both Charles Darwin and Charles Darwin's earthworms (1881) were capable of making intelligent decisions. The difference is that the earthworms lacked consciousness but displayed what Daniel Dennett has called 'competence without comprehension' (2018). It is also as if evolution itself shows competence without comprehension, as its positive selections are, by definition, the survivors that breed.

The first half of the twentieth century was the twilight of a science era when everything had to be counted, including human intelligence. Psychologists became obsessed with the reductionist dream of a finding a single number – a quotient – whose magnitude indicated the intelligent capacity of a single person. The intelligence quotient, IQ, was pursued with the same enthusiasm that a previous generation had sought the missing link and was even more evasive. Attempts to quantify intelligence led to the development of an intelligence test of routine questions. The person of average intelligence scored 100.

Intelligence was also supposed to be heritable. This view was mainly due to Sir Cyril Burt, who seems to have manipulated his results in order to support his personal prejudices. Burt, as a young man, was impressed by Francis Galton and his influential eugenics movement. An early paper by Burt (1909) described the results obtained when the same intelligence test was applied to upper-class children in private preparatory schools and to lower-class children in elementary schools. The former did better and, no doubt much to the delight of the Edwardian Establishment, he concluded that the differences were innate. This implied, falsely, that the upper class was therefore fitter, in a Darwinian sense, while the lower classes deserved their various fates.

After Burt died, his work was re-examined by others. His studies of the intelligence of identical twins that had been separated and brought up in different homes suggested that, irrespective of nurture, their genetic similarity ensured they did equally well in the tests. Separated twins are difficult to come by, and Burt used an unreasonable number of pairs in his work. Efforts to trace many of them have failed. Two of his co-workers are similarly untraceable. The jury is out: did he falsify his studies or, like others before him, was he 'helped' by well-meaning assistants? Did he *have* any assistants? It is still a contentious issue. The IQ tests of 70 years ago still find occasional use but it is agreed that whatever it is that they measure is only a tiny part of human intelligence.

In discussing evolution by natural selection, the most appropriate definitions of intelligence come not from studies of human intelligence but from the field of artificial intelligence. Eighteen such are quoted in the Legg and Hutter paper mentioned above. All are relevant to different extents but two resonate. They are:

> Any system … that generates adaptive behaviour to meet goals in a range of environments can be said to be intelligent.

> Intelligence means getting better over time.

According to these definitions, evolution proceeds as if it had both 'intelligence' and 'purpose'. It also has 'memory', in highly conserved portions of DNA shared by every living thing, while every individual organism it creates displays both 'intelligence' and 'purpose' (competence without comprehension; staying alive). Higher animals and humanity display a form of competence *with* comprehension, which we call consciousness.

Natural selection seems to lead to increased complexity. The 'Galaxy Song' that is sung by Eric Idle as he steps out of the refrigerator in the Monty Python film *The Meaning of Life* contains the phrase 'how amazingly unlikely is your birth'. That particular likelihood depends entirely on where you want to start. If we claim poetic licence and consider *birth* to refer to your ultimate origin, perhaps the very first cell ever, many people would now say that was a very likely event, given the circumstances in which it is imagined to have arisen. It is an event that has had consequences 3 or 4 billion years down the track, one of which is you. It now transpires, from a variety of sources of evidence, that given the conditions that existed on the primitive Earth, the origin of life was almost inevitable, barring cosmic accidents. The odds favoured the origin of life. It was a racing certainty.

Climbing aboard this idea, one can travel in one's mind back down the genealogy of life to our very first hypothetical ancestors. This ancient type of cell, of which many representatives are with us today as bacteria, blue-green algae and many other highly specialised forms, was tiny, with its hereditary material dispersed in its cell substance. It is now accepted today that the first of the 'modern' cells – the ones of which *we* are made – was put together from ancient cells; three of them, in the case of animals, or four, in the case of plants. It should be clearly understood that so-called modern cells are pretty ancient, too, perhaps more than 3 billion years old, just less so than their 4-billion-year-old predecessors! This idea was the great contribution made by Lynn Margulis (1970). As a consequence, the modern cells are very different from the 'ancient' bacteria-like ones, although trillions of the latter are still with us in the world today.

Here is the cast of players, all ancient cells (Figure 6a):

Figure 6a. The ancient cells.

First is the ancient host cell. Little more than a sluggishly moving blob, it is anaerobic, using something other than oxygen for inefficient energy production. It probably lived on any ancient cell it managed to trap and engulf. Next come the oxygen respirers, shown as little cigar-shaped cells, the survivors of the great oxygen pollution event brought about by the blue-green algal bloom that introduced oxygen into the atmosphere. The blue-greens are represented in Figure 6a as a green oval. Finally, and most enigmatic, are the spirochaete-like cells – twisted threads that had developed intracellular 'muscles', called microtubules, that enable them to move about, and to move things *inside* themselves.

Figure 6b. Ancient cells transforming into modern animal cells, step by step.

The sequence of events cannot be known for certain, but it is plausible to think that an early step was the ancient host coming to terms with the oxygen respirer, creating an enhanced ancient cell capable of respiring oxygen (Figure 6b, upper right). The next step (centre) is likely to have been the acquisition of the spirochaete-like cell to equip the new partnership with microtubules and to create an internal mechanism that physically pulls the clump of DNA in two parts for accurate – but not too accurate – cell division. That step also contributed the cilia and flagella that allow the whole cell to move in its environment. Each cilium and flagellum in a modern cell has its own basal granule of DNA and there is DNA associated with the cell division mechanism, the centriole. Finally, the further acquisition one or more blue-green algae (Figure 6a, extreme right), and their capacity for photosynthesis, gave birth to the plant kingdom.

Figure 7. Modern cells: single-celled animals and plants.

After, no doubt, millions of years of evolution, the result (Figure 7) was something like the single-celled animal (left) or plant (right) that we know today, with its DNA wrapped up in a nucleus (the white circle).

How can we possibly know that this is true, you ask? Well, of course, we can't, it's a just-so story to help us understand what happened. *Something* like it must have happened and there are organisms with other inclusions, such as hydrogenosomes in parasitic protozoa and the apicoplasts in the malaria parasite and its relatives, that seem to have been derived by symbiosis with different ancient cells in a similar way. The world of ancient cells must have been a maelstrom of symbiotic experiment.

This DNA is the modern cell's genetic program in coded form, which carries the information necessary to make another cell. Modern cells also carry, outside the nucleus, but within the cytoplasm, DNA derived from those early symbioses; from chloroplasts and mitochondria and, in centrioles and the basal granules of cilia and flagella, from the spirochaete-like ancient cells. These small bodies inside the cell are what is left of the ancient cells that went to make up modern cells.

Every living thing on Earth, other than the ancient cells, is made up of modern cells and, therefore, from the ancient cells onward, every living thing is related to every other living thing. The whole of the biosphere (all life on the planet) was implicit in that very first cell. Inside those ancient cells, the DNA code used for making a new cell is more or less the same as that in modern cells.

Once the first modern cell had come into existence, and assuming no pan-global catastrophe occurred to wipe it out, an evolving biosphere was assured. The form that the biosphere took was contingent on circumstances. Life might simply have remained a covering of bacteria-like cells on a stony ball but instead it took the next step and gave birth to the modern cell. Once that happened, and once the first modern cells got the trick of working together, it was inevitable that life would increase in complexity and size with time.

And the triumphant result after 1 million or so years of experiment is the 'modern' cell (pink) that is the ancestor of all animal and fungal cells and the 'modern' cell (green) that is the ancestor of all plant cells (Figure 7). The white blob represents the nucleus, as coordinating the DNA from three or four cooperating symbionts requires a central organiser.

Complexity in a living organism implies structure. Structure requires the information necessary to build it. As complexity increased, the amount of structural information associated with the biosphere also increased. As intelligence is founded on the application of information to solving the problems of survival, the quality and amount of intelligence also increased. In fact, evolution acts as if it were intelligent even if it is not in a form immediately apparent to human beings.

Humans too are subject to natural selection. Selection occurs for genes for human hair type, skin colour or body shape, or for forming social groups under changing geographic conditions, such as the ending of an ice age. During the last ice age, natural selection allowed the appearance in the north of human variants who had lighter-coloured skins. It was probably because light skin permitted more UV light to pass through for vitamin D synthesis in these less sunny northern latitudes. At the equator, a darker skin was more advantageous, as it acted as a sunscreen, limiting UV damage while still permitting the passage of enough UV light for vitamin D formation.

Since the environment does not remain constant, organisms must also change to keep up. The current rate of global warming is set to provide humanity with an object lesson in the truth of this statement. The more an environment changes, the greater is the selection pressure exerted on its inhabitants.

Unstable environments, therefore, or the margins of stable environments, are hotbeds of evolution. The wildflowers of Western Australia are an illustration of this phenomenon. In Kalbarri National Park are dozens of flower species unique to its impoverished sandy soils and hot summers. The rate of speciation there smacks almost of desperation!

Human societies are part of, not separate from, this process. Over generations, wars have brought partnerships between the victors and the defeated in various combinations, giving the gene pool a vigorous stir. The Vikings, once at war with Britain, became part of Britain and contributed a Viking strand in the northern British gene pool. The mass population movements of the twenty-first century are a continuation of the migrations of peoples that began at least 80,000 years ago out of Africa, all contributing to human evolutionary change. The future of human variation rests on the major flows of populations across the globe.

It is obvious that all forms of life must achieve a close relationship with their physical and social environments. They behave in an intelligent manner to establish it and are competent to ensure that this relationship continues. However, as any animal lover will tell you, the more complex plants and animals display something that is more than competence alone.

Consider the well-known single-celled 'slipper animalcule', the pond-dwelling *Paramecium*. This tiny freshwater organism has some well-developed avoidance mechanisms for getting itself out of a fix when it bumps into one. At the same time, it has to regulate its internal environment by getting rid of metabolic wastes and pumping out excess water through its vacuoles. In fact, it is ensuring a continual internal steady state, and because of this, stays alive. This surely is 'existential intelligence', arising from direct response to the environment, even if not accompanied by consciousness.

Charles Darwin (1881) certainly felt that his earthworms could be described as intelligent. Darwin's colleagues considered his earthworm study as a trifle for his old age. Yet his work showed that the entire surface of England rises significantly due to the continual deposit of earthworm casts. He wrote of earthworms choosing the right-sized leaves to haul into their burrows, without trial and error. He wrote:

> If worms have the power of acquiring some notion, however rude,
> of the shape of an object and over their burrows, as seems the case,
> they deserve to be called intelligent, for they act in nearly the same
> manner as would man under similar circumstances.

Even a sedentary organism, one that stays put, such as, say, ivy, must be credited with existential intelligence. Its roots are continually exploring the earth in search of nutrients and water; its tendrils are continually probing, seeking places for attachment; its leaves are continually adjusting themselves to maximise their access to light.

Philosophers, such as Midgely (1980, 2004) and Ruse (2013) now allow us to assume that many of the higher animals experience a form of consciousness. It therefore would be churlish to deny them forms of intelligence. The spatial intelligence of birds allows them to live in a world where their capacity for flight gives them access to an extended third dimension that humans only dreamed about until the beginning of the last century. Spatial intelligence has to be extremely well developed in an animal that moves freely in three dimensions (fish, birds, whales etc.). Perhaps it is even qualitatively different from that encountered in ourselves and our primate cousins. Humans are effectively confined to two dimensions of the Earth's surface, if you discount walking up and down hills and climbing trees.

If you doubt that spatial intelligence is highly developed in birds, join the remarkable flight of Darshan the eagle on YouTube (at: www.youtube.com/watch?v=um8M9azpmb4), as he is launched by a handler from the highest building in Dubai to search, successfully, for his trainer on the ground. Marvel at the aerial stability of his body as a platform for an ever-seeking head – and, in this case, a camera to record the experience. Spatial intelligence must also be well developed in cetaceans. Think of the global migrations of blue whales or the vertical expeditions of sperm whales into the deeps for their prey. On top of that, try to imagine the effect that sheer size has on bodily-kinaesthetic intelligence. These animals are so removed from us in evolutionary space that we can only begin to speculate on the workings of their brains.

A dog is a supremely olfactory animal, living in a world where scents, important reference points, swirl about it in a way that we would find bewildering. We can only begin to understand, with our own rather poor olfaction, a dog's way of knowing. For example, it is a spring morning outside, and a warm breeze brings a whiff of floral perfume. We can sense the direction from which the breeze comes and also identify the flower as lilac. But what else is on that breeze? Jip, the sentient dog genius of Hugh Lofting's (1926) famous creation, Dr Dolittle, might tell you! If he were specially gifted, he might have a nose as much as 100,000 times more sensitive than yours; the average pooch can certainly manage 10,000 times. Tales of the dog's acuity of smell are many. For example, some gifted dogs can sense killer whale faeces a mile away, although we hesitate to ask where lies the advantage. There are companion dogs that sniff out ketones and warn a diabetic person that an attack is imminent. Diagnosticians are greatly interested in cancer-sniffing dogs. How does a dog's brain process this information? These ways of knowing the world are all equally valid in the quest for survival. We are all here, sharing the same world, but we relate in different ways to different aspects of it.

Humans are able to use their minds in various ways to learn about the world and to respond to it. We are of course limited by our physical senses, compared with some animals, and by the information processors with which evolution has equipped us. We are, however, the great generalists. Our sense organs pick up information from a great many sources. We are blessed with the processing equipment that gives us a greater chance than most of integrating the different sorts of information supplied by our senses into a view of the world that we can live in. But this is only one view among many.

If we have difficulty understanding a dog's way of knowing, an animal with which we have about 82 per cent DNA similarity, how much harder must it be with a bird? The similarity between a mammal and a bird is about 60 per cent, further complicated by different gene organisation and a different brain architecture in the bird that crams many more cells into a small space. Mammals and birds have followed different genetic destinies for at least 200 million years. Although bird brains seem relatively small, some of them – such as the intensely studied, clever, tool-using crows from Caledonia – have at least the processing capacity of the great apes. Their assemblages of sense organs and ways of knowing are very different from those of mammals.

Homing pigeons and migratory birds navigate by sensory signals we are only just beginning to understand; among them is an inbuilt magnetic compass. Birds have bodily-kinaesthetic and spatial understandings for flight and homing. One has only to watch the gyrations of a restless flycatcher on the hunt for flying insects, or the total stillness of a white-faced heron fishing, or the untiring flight of the albatross on cleverly locked outspread wings as it skims the aerial up-currents caused by ocean waves, to know that. Swifts apparently sleep on the wing, half a brain at a time. What does that feel like?

According to studies of human ways of thinking, humans have at least nine ways of knowing their physical and social environments. Howard Gardiner has identified eight forms of human intelligence: inter and intrapersonal, logico-mathematical, bodily-kinaesthetic, musical-rhythmic, visual-spatial, verbal-linguistic and naturalistic. Brown (2008), and Brown and Harris (2014), have also identified the ways in which our social environment has shaped our thinking: as individuals, as community members, as specialists, as members of organisation organisations and as 'wholists' and more. Animals, collectively, must also have many different 'ways of knowing'.

We have already accorded plants an existential intelligence. Plants, however, have many other strategies, than the few we discussed, that all contribute to the great and intelligent aim of staying alive. Pitcher plants drown insects in leaves modified as containers, to obtain their nutrients. Sundews slowly enclose their prey with sticky tentacles. Trees, when damaged, release into the air molecules of substances that alert other trees in the vicinity to the presence of browsers, inducing them to produce substances that make them less palatable.

Gaia has acted with intelligence, with competence but without comprehension, to ensure her continued survival for more than 4 billion years. With the evolutionary advent of animals, and humans in particular, she can now add comprehension to her vast list of connected attributes. In *The Hitchhiker's Guide to the Galaxy*, Douglas Adams posits that the Earth is a giant computer, run by mice, a computer that has been crunching away at a problem for millennia. Can there be a sliver of truth for we humans in this fanciful image? Perhaps yes, if you consider how we take for granted the mechanisms by which Gaia has solved the question of continuous existence.

Gaia is essentially a self-regulating system that acts as if its objective were its own continued existence. Over 4 billion years, the golden thread of cooperative evolution has tended towards increased complexity. This is not to say that the simpler life forms that were created along the way have been discarded. If they have withstood the tests of natural selection, they are still with us. Thus, much of the world belongs to ancient cells, and the blue-green algae have been and still are important in shaping it by pouring out their waste product, oxygen. The first of the modern cells embarked with great success on the voyage towards multicellular complexity, but they also explored the capacity of such cells to achieve complexity within a single boundary membrane. This includes the great kingdom of the Protista, modern single-celled organisms, illustrating what can be achieved even at that scale.

Similarly, we see by looking around us the successful steps towards multicellular complexity. Sponges are aggregations of cooperating cells that still have a high level of individuality. They will re-aggregate if separated. The Volvocales (Figure 8) are spherical structures made up of cooperating unicellular algae. Obviously, it is a successful way of life as clusters stay together after a fixed number of divisions – up to *Volvox* itself, a construct of about 16 accumulating divisions. And successful because they are still here, as are liverworts, mosses, ferns and higher plants.

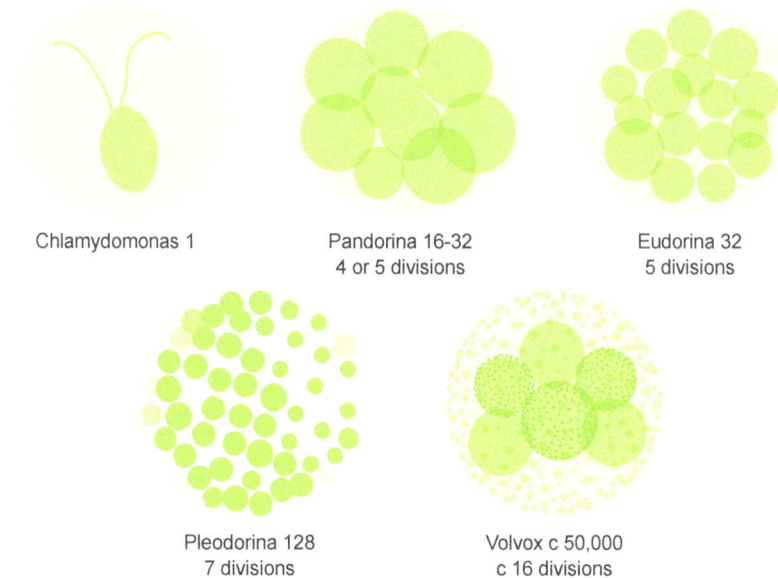

Chlamydomonas 1

Pandorina 16-32
4 or 5 divisions

Eudorina 32
5 divisions

Pleodorina 128
7 divisions

Volvox c 50,000
c 16 divisions

Figure 8. The amazing Volvocales.

There seems, however, to be another possibility offered by increased complexity, and that is the evolution of intelligence at high levels of functionality. This interesting phenomenon has arisen separately in a number of groups; in molluscs (octopuses), birds (in particular, in the crow and parrot families) and in mammals. In the course of evolution, brains have come to represent the highest level of complexity within Gaia. There are trillions of connections within a human brain.

An important characteristic of natural selection is that it has 'memory'. The reason we can say that all living things are related to some degree is that clusters of genes that have worked well in the past are conserved in the present. There are various degrees of conservation; highly conserved sequences are associated with vital functions like cellular respiration. The human foetus *in utero* passes through a stage where gill rudiments are formed; after all, they share 70 per cent of their genes with zebra fish. That first cry indicates the baby's switch to breathing air in their new environment. Even tomatoes share a great number of genes with humans, the ones that look after essential cellular processes. And as someone once said, 'being human is just another way of being a fish'! But so is being a bird.

One of the many great surprises afforded by modern genetics was the discovery that the human genome comprised less than a quarter of an expected 100,000-plus genes. It turns out that this is possible because no gene acts in isolation, but only in combination with other genes. It is as if individual genes are like individual neurons that work in combination with other neurons if we are to walk and talk. A particular characteristic – let us say, the possession of teeth – is represented by several genes, that make a sort of gene cluster. If, during evolution, teeth are no longer needed, their gene cluster does not get reinforced and, with time and the absence of selection pressure, may disappear altogether. But sometimes this cluster that is no longer useful persists among all the other apparently unused bits of DNA. Ancestral birds had teeth, but modern ones have beaks. If something goes awry in development, a few teeth may appear because the relevant gene cluster is still there and has been awakened by something going slightly wrong with the way the genome is translated (the professional term is 'expressed') into flesh and blood and bone. The saying 'rare as hen's teeth' has power because our experience is that hen's teeth are indeed uncommon but do crop up from time to time. And it is interesting that when they do, they appear to be quite crocodilian, a glimpse into their reptilian past (Harris et al. 2006).

The genetic sequence for hen's teeth is an example of a gene cluster that is not being conserved in birds but is occasionally expressed accidentally. When, however, a cluster of genes codes for a really important function, such as the ability to manipulate and store energy, those genes are selected together, and are conserved – 'remembered' – for use in subsequent generations. There are many such highly conserved portions of the genome – that is why we have part of our DNA in common with tomato plants.

While very useful gene clusters are preserved in the genome upon successful Darwinian selection, all the other gene clusters that have proved useful fade without continual reinforcement. The discovery that the 'eye cluster' from chickens works in the fruit fly, *Drosophila*, was quite startling because chickens and fruit flies have not enjoyed a common ancestor for more than half a billion years (Tomarev et al. 1996). Even so, fruit fly–like eyes were produced rather than chicken eyes when the chicken genes were introduced into the early developmental stages of the fruit fly. It appears that the gene cluster is, effectively, a command that orders 'make an eye'. The response is to make an eye of the type determined by other surrounding genes: the context in which the eye cluster finds itself. It is all about context; genes act in teams and the local genetic environment of the activated gene is of the utmost importance.

The *Oxford English Dictionary* (1993) defines connectivity as the state, property or degree of being interconnected. The number of connections being made within Gaia at any moment is astronomical. Think of all of the animal offspring, of all the plant seeds, of all the spores of mosses and ferns and fungi, like dust, flying in the wind. Add in the reproductive adventures of the bacteria and Protista, and fish and whales, and everything that lives in water, and elephants and people and everything that lives on land, and you have an image of the biosphere, of Gaia, as a great engine of connectivity and natural selection. In fact, it resembles a great computer but, as far as we know, it is not run by mice and nor is the answer 42!

A very important property of connectivity, especially connectivity on a large scale, is that it leads to unexpected outcomes – emergent consequences. Here is a thought experiment that you can do, devised by Stanley Kauffman (1995).

Imagine a scatter of several hundred similar buttons on a flat surface, distributed so that no one button is touching another. Select two buttons randomly and join them with a thread and replace them on the table. Repeat the process and go on doing it. For the first few tries, it is unlikely that you will pick up a button that has already been joined to another one. As more and more buttons are joined, however it becomes increasingly likely that at the next try you will pick up, say, an unjoined button and a button that has already been joined to another. Make a threesome. If, next time, you pick up two strings of two buttons, make a foursome. Soon you will be creating fives and sixes and so on. Continue in this way and eventually each button you select will be attached to every other button. A net has been formed. And a net has properties – it can be made to hold things, for example – that a few handfuls of unconnected buttons and strings does not.

One can imagine evolution proceeding in this way. The ancient cell appears and proliferates. Three or four quite different ancient cells integrate their capabilities to form a modern cell. Cooperative evolution begins! The modern cell has a suite of advantages not possessed by its single-celled forebears. It proliferates into a wide range of environments (adaptive radiation). Some modern cells get together and form simple multicellular plants or animals … And so on, right up to the levels of simple ecosystems and beyond. At each stage, the emergent consequences set the scene for another round of adaptive radiation.

A recent paper explored the role of connectivity in understanding how evolution and ecology work together. The authors, Watson and Szathmary, suggested that, even though an ecological system was not a Darwinian unit for natural selection, such a system and all its components can behave like a learning system, promoting its adaptation to environmental conditions. It has even been suggested that landscapes – another name for an interconnected group of ecosystems – can adapt. A study of landscape recovery after clearing by humans or by fire showed that the succession of vegetation regrowth after clearance was not consistent, as one might have been led to expect in final-year botany. It depended very much on what had gone on before, the previous history of the system.

Another way that landscapes can evolve is by interacting with their animal occupants. The grassy South Downs of England owe their existence to browsing by rabbits, introduced 1,000 years ago, and grazing by sheep. The American prairies were maintained by tens of millions of bison. This is niche improvement on a large scale. Niche improvement occurs when animals and plants interact with their environment in such a way as to improve their chances of survival, thus increasing their Darwinian fitness. They change the niche as well as having to change themselves. Animals build nests, mounds and hives, and plants drop leaves that are converted to nutrient humus by fungal and bacterial activity. Trees can change the water table and water flow. By their activities, living organisms influence the way the environment changes.

The biosphere is thus continually changing in a coordinated manner as each of its components undergoes cooperative evolution. Many other things than genes are transmitted from parent to offspring: components of the egg, hormones, compounds that can switch genes on and off, antibodies and learned knowledge. Further, a given set of genes has the power to produce different types of adults. For example, the sex of some reptiles is determined by temperature: if the nest temperature is above 32°C, mainly males are produced. Seal point cats are fawn with brown extremities – tips of ears and tail, paws and nose. These are the coolest parts of the body, where the cold activates a gene that produces pigment that can absorb heat from the Sun.

Humans are particularly good at improving their environment to enhance their own survival and are able to live in communities from the polar ice to the equator. As a survival technique, however, it carries its own risks. The human release of fossil fuels is threatening to change the currently human-friendly atmosphere and there has been too little time for a genetic response to counter the threat, while a concerted social response is a victim of political inertia.

Human risk perception is poorly adapted to long-term dangers. It took more than half a century to convince people that smoking causes lung disease and cancer; even now it is a risk that many choose to take. And now, for human societies, studies of changes induced by the interaction of social systems with the human genome are only just beginning. All of these things have the effect of channelling evolution and providing, in human offspring, material for natural selection.

According to several schools of thought, an individual human intelligence is made up of contributions from a palette of different qualities of mind. They are listed here, in no special order of importance:

Table 1. Human qualities of mind.

1. Intelligences
Naturalist; Musical; Logical-mathematical; Existential; Interpersonal
Bodily-kinaesthetic; Linguistic; Intrapersonal; Spatial
2. Ways of living
Disciplined; Synthesising; Creative; Ethical; Respectful
3. Problem-solving
4. Collective thinking
Physical; Societal; Ethical; Aesthetic; Sympathetic
5. Knowledge cultures
Personal; Community; Specialised; Organisational; Holistic

Sources: Gardner (1983, 2007); De Bono (1999); Brown (2008); Brown and Harris (2014).

Multiple ways of thinking, intelligences, are mixed in varying proportions in every human being. In this way, they offer a very large number of possible combinations, which allow humans to think collectively in a spectrum all the way to genius. There are, however, questions that are usually, and quite rightly in our opinion, left unanswered: for example, how many, and how much of each category is required to make a functional human being? It is sufficient to say that the mix of intelligence, emotional understanding and creative thinking has permitted humans to be resilient and adaptive.

Each of these ways of thinking contributes to the great objective of survival that humans share with every living being. All living organisms are dedicated to the universal enterprise of staying alive, and to do this they must act with intelligence, either unconsciously or consciously. The logical and empirical capacity in our thinking, that we call human intelligence, is obviously only one way of understanding the world. When talking about non-sentient creatures, it may help to think in terms of 'ways of knowing'.

4

HOW EVOLUTION WORKS

In which we explore some of the many mechanisms of evolutionary change.

The engine that drives the evolutionary train forward is Darwinian – or natural – selection. It is fuelled by the continual changes made by living organisms in their efforts to complete full and successful lives, so that their genes are represented in successive generations. This chapter explores some (only some – a full list would make a book of its own) of the many ways by which this happens, thereby providing the material for selection. At every stage of evolution innovation creates more opportunities for cooperation, and emergent forms in the shapes of new organisms and fresh collaborations.

Offspring are subtly different from their parents, and all offspring are subject to the process of natural selection at every stage in their lives. At first, it was thought that the main source of variation was the spontaneous change to DNA that occurred as the result of faulty copying, or that were brought about by external factors such as radiation. Collectively, these changes were called mutations.

Since then, there have been major advances in understanding the interactions between organisms and their environments and the exchanges that take place spontaneously between different sections of DNA. Some of these are now being exploited in medicine by gene therapists. For example, the replacement or repair of faulty genes in embryos, or modifying the patient's own blood cells to become more efficient cancer fighters would have once been considered miracles. The new discoveries have brought a far wider appreciation of the sources of, and the persistence of, mutations. In this chapter, we explore a range of these sources of change. Remember, a single gene may have different effects, depending on the other genes in its immediate environment. Nothing so much reinforces da Vinci's insistence that everything is connected to everything else as apparently random changes that are not random after all.

To the surprise – and even alarm – of his contemporaries, Darwin included our own species, *Homo sapiens*, in his discoveries of evolutionary change. In this chapter, therefore, all the phenomena described may also apply to us. To this rich biological inheritance for humans, however, must be added our evolution as a social species, our artistic nature and our capacity for empathy for others. Biological evolutionary pressures, such as our long generation time and its implication for extended parental care, contribute to the social changes.

The accumulation of small mutations leading to new species was attractive to the gradualist point of view of evolution, first espoused by Darwin himself. It was a view that held sway in the first half of the twentieth century. Its great flaw, as Richard Goldschmidt (1940) saw, was that it did not provide a mechanism for macro-evolution – evolution *beyond* the family level of classification – although it did explain very nicely how families might arise. We all of us recognise the members of the family of parrots but, with all due deference to psittacophiles, on first observation one parrot looks very like another. Few people would have difficulty believing that a palm cockatoo and a sulphur-crested cockatoo are related, but a crested pigeon is something else entirely. How do you get a pigeon? It is counterintuitive that chihuahuas and borzois are the same species of dog, but chimpanzees, gorillas, baboons and monkeys are easily recognised as a group.

Goldschmidt (1940) suggested that the genetic system might make a number of simultaneous changes in a single generation to produce what, unfortunately, became called 'hopeful monsters'. In vain, Goldschmidt pointed out that he did not mean that the individual monster 'hoping' for its own survival would be fully adapted to its environment but that all of its variants would be subject to the pressures of selection. Unfortunately, the reductionist 'either/or' attitude prevailed. No-one was prepared to countenance the idea that one might have both mechanisms – mutation *and* shuffling of genetic material without mutation – at the same time.

In humans, social acceptance is strongly influenced by extreme variation, for facial distortions in particular. Mutations, such as albinism or dwarfism were often treated to the kind of extreme discrimination well understood by many modern human minorities. 'Freak shows' at circuses or public display in asylums have been the sad fate of so many in the past. All these responses contribute to human social evolutionary pressures so the concept of the 'hopeful monster' was doomed. Now we know, however, that genes often act in teams and the laborious, one at a time, accumulation of single mutations is only part of the story (see Chouard 2010).

One of the more important and common ways that one species might split into two is genetic drift. Imagine a population of mice living on a peninsula that is attached to the mainland by a narrow neck of land. Every mouse shares the same genetic heritage with every other mouse on the peninsula and on the adjacent mainland. Each male mouse therefore has the potential to interbreed with every female mouse. There is always the possibility of the flow of genetic information back and forth, as A breeds with B, and B breeds with C, and C with A and so on. Now imagine a minor catastrophe – the sea erodes away the link with the mainland. It separates the mouse population into two, one on the mainland and the other on the newly formed island. Now gene transfer through the whole population cannot occur. Mainland and island mice can never get together. Each separate population is then subject to different environmental conditions that, over a number of generations, affect its genetic heritage and become reflected in its members. The two populations begin to diverge and embark on their own evolutionary adventures. Eventually they will become distinct subspecies and, finally, distinct species – especially if the island lacks the predators that are found on the mainland. The popular holiday resort of Lord Howe Island is a good example; its isolation has permitted the evolution of many species that can only be found there.

The same is true of the separation of human populations in the last ice age. Cut off in the frozen north, Siberian adaptations include increased activity of two genes that burn fat for heat rather than energy and others that are implicated in the metabolism of fats from meat and dairy products (Cardona et al. 2014). Outer manifestations include eyes with epicanthic folds to shield them from the icy wind, and facial sinuses with thick fat pads to protect them from freezing conditions.

Islands are marvellous evolutionary laboratories. The extinct dodo, a flightless bird that once lived on Mauritius and stood a metre tall, showed what pigeons were capable of becoming, given enough time and the absence of predators (Quammen 1996). Even quite large isolated islands, such as New Zealand, had their own unique inhabitants. The giant moas, flightless birds almost 4 metres high, persisted until the arrival of Māori, about 600 years ago. Sadly, both the dodo and the moa were hunted to extinction.

These examples, to which we can add the giant turtles of the Galapagos and the Komodo dragons (giant lizards) of Indonesia, represent one response to isolation. Another is dwarfism. The limited resources on islands sometimes result in dwarfism as an evolutionary outcome. Limited resources create a selection pressure towards achieving reproductive ability at the smallest size consistent with long-term survival. Humans are not exempt from these pressures. Recently, on the island of Flores, the discovery of bones of *Homo floresiensis* prompted a rethink of human history. These small hominins were just over a metre tall. They lived alongside dwarf elephants, which had also undergone a similar process (Aiello 2010). The small humans were nicknamed 'hobbits' after JRR Tolkien's creations.

Another opportunity for variation is offered by increasing the amount of DNA in the nucleus of the cell. Doubling or tripling the number of copies, up to 12 times, is not uncommon. The more DNA an organism possesses, the greater are the chances that it will vary between generations and the greater are the opportunities for mutation to occur. The phenomenon is called polyploidy and often occurs in nature. It is frequently observed in our domestic crops after the thousands of years of selective breeding they have undergone. The common black and white Australian mudlark appears to be a tetraploid (doubled) version of the much smaller diploid African and Asian common pied flycatcher. The common dandelion (*Taraxacum*) is also tetraploid. Doubling of individual chromosomes is common among humans, but usually leads to genetic disorders of varying severity.

The genetic program is a dynamic system, bits of which are continually changing places and interacting, creating new patterns. Nothing illustrates this better than the jumping genes (transposons) discovered by Barbara McClintock in the 1940s and 50s. 'Transposition mutagenesis' allows genes to be transferred from one chromosome to another, thus interrupting or modifying the functions of their new neighbouring genes. McClintock was an example of a paradigm shifter who was ridiculed by her contemporaries. The importance of her work was finally acknowledged with the Nobel Prize in 1982 (Keller 1983).

Meiotic drive is the name for another interesting phenomenon that illustrates the dynamic character of the genome (Sandler and Novitski 1957). It is a complicated mechanism that is sometimes encountered in some species during cell division. It results in genes from the male or female gamete being over-represented in the fertilised egg. In other words, contrary to classical Mendelian genetics, where the 50:50 ratio of gene distribution is to be expected, it is skewed in favour of certain genes. It is as if, here, we have an example of evolution evolving, bootstrapping itself, by increasing the chances of the 'favoured' gene being represented in successive populations.

Now we come to one of the most important creators of genomic novelty, that of horizontal gene transfer. The original Darwinian view, one that persisted until the middle of the last century, was that vertical transmission of genes, from parent to offspring, was the only game in town. This provided the metaphor of the evolutionary tree or bush. We know now that it is certainly not the only game and that it never was. The whole of life other than bacterial is based on an initial collaborative event in which three or four independent sets of genes from bacteria-like organisms merged their identities in a single cell. This is horizontal transfer, not branching. The evolutionary tree, with one main trunk and lots of side branches is therefore not an accurate metaphor for evolution, at either the cellular or the primate levels of classification.

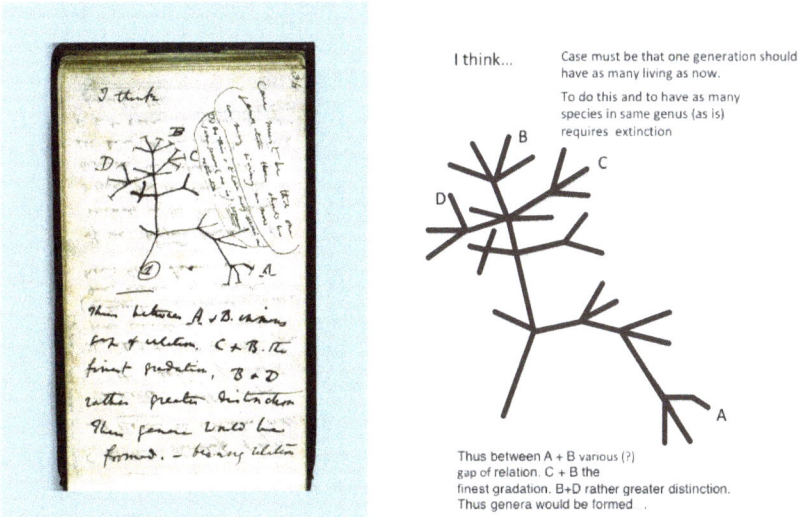

I think...

Case must be that one generation should have as many living as now.

To do this and to have as many species in same genus (as is) requires extinction

Thus between A + B various (?) gap of relation. C + B the finest gradation. B+D rather greater distinction. Thus genera would be formed .

Figure 9. Darwin's very first evolutionary map.

Darwin did not think of a tree in the first place. His sketch of a possible evolutionary pattern is more of a map (Figure 9). The evolutionary bush, with many main stems and lots of twigs, is better, but still unsatisfactory. Best of all would be a three-dimensional representation of an evolutionary network that shows how adjacent and even more distant branches of the evolutionary bush are joined by horizontal connections.

The first great act of horizontal gene transfer to produce modern cells had proved to be an enormous evolutionary success. On the principle that what worked once can work again, which seems to pervade the evolutionary story, it is not surprising to discover that horizontal gene transfer from one cell to another still occurs commonly among ancient cells and, for example, is responsible for the development of resistance to antibiotics in pathogenic bacteria.

Viruses also have a role. They are not cells, but some may once have been cells. Now they are packages of DNA or RNA that depend on cells to reproduce. It is hardly surprising, therefore, that viral inclusions in different genomes are common. This is never so clear as when viruses cross from animals to humans. A recent and dangerous example is Severe Acute Respiratory Syndrome (SARS), a virulent, invasive respiratory disease that startled the world in 2002–04 with an overall death rate of about 15 per cent, although some sections of the community suffered death rates as high as 50 per cent. It required full international collaboration to identify the source of the virus as the Chinese horseshoe bat. The bats infected civets, a wild, cat-like animal related to stoats and weasels. When they were brought together in Chinese food markets, the viruses transferred to humans. At the time of writing, we are in the grip of a pandemic, COVID-19, due to another virus that has made a similar jump.

If the host cells survive the viral attack, some viral nucleic acid sequences may persist in the cell, silent at first but, with time, they may be incorporated into the genome. In this event, they augment the genome and have the potential to become involved in natural selection. In humans it is currently thought that about 8 per cent of our genome is of viral origin. The fact that foreign gene insertion into a genome is a relatively easy outcome to achieve is demonstrated by the work in many laboratories, such as those striving for blue roses, or enhanced vegetable oil yields in rape plants, and the many other transgenic food crops now available. Another application is therapeutic gene transfer. Genes can be added to human cells or removed from them. Leukemia is one condition that responds to the replacement of malfunctioning genes by healthy units from the patient's own genetic repertoire.

Another quite remarkable use of gene technology is known as CRISPR (pronounce it 'crisper') and it is an instantly memorable acronym for the uncatchy 'Clustered Regularly Interspaced Short Palindromic Repeats'. It originates as a defence mechanism found in bacteria where its function is to identify the DNA of an invading virus and inactivate it. Modified further in the laboratory, it is now being tested for use in embryos, for identifying and removing deleterious genes or introducing beneficial ones. Potential targets include diseases such muscular dystrophy, congenital blindness, haemophilia, Huntington's disease, cystic fibrosis and a range of cancers.

A more detailed explanation of CRISPR is to be found on the Internet at: www.wired.co.uk/article/crispr-cas9-technique-explained.

Having looked at some of the ways that a single genome can be shuffled for the next generation, what are the possibilities for two or more genomes working together? The cells of which we are all made, that are formed from collaborating bacteria-like organisms, come once again to mind. Evolution, nothing if not iterative, has tried the experiment of collaboration over and over again. At the whole organism level, wherever you look there are successful associations, such as lichens, that are an ancient partnership between fungi and ancient (blue-green) algae or modern algae, right up to the modern ruminants that could not survive without the fermenters – fungi, protozoa and bacteria – in their rumens. The result is that Gaia is made up of countless beneficial biotic associations. In addition, every known organism has at least one parasite, whether viral, bacterial or modern cell. A quick consideration of this proposition should convince you of the thesis that one of half the living world is therefore living in some sort of association with the other and that the numbers of such associations are astronomical. These associations can be graded.

At the first level, there is predation, as exhibited by lions hunting springboks. Next come the things that live on the outside of their hosts and plunder them in some way. These are the ectoparasites, such as leeches and fleas on mammals, and green- and black-fly on plants. They are 'predators' too: the difference is in the relative sizes of the food source and the predator. This is acknowledged in that we talk about the lion's 'prey' and the flea's 'host'. Humans regard fleas as pests, not life-threatening unless they carry plague. Though, fortunately, we do not often form the prey of lions, we are often hosts to many small ectoparasites like fleas. From a springbok's point of view, a lion is more than just a pest; it is indeed a predator.

It is when the pest starts eating the prey from the *inside* that the game changes. This is true parasitism, and all sorts of genetic changes are required of the internal predator or parasite if the parasitic relationship is to exist and perhaps evolve into something mutually beneficial. This is because the host fights back, perhaps by mounting a strong immune response against its invader. That, in turn, calls for a secondary response from the parasite, and so on. This is an example, common in biology, of the phenomenon that was called 'Running the Red Queen's Race' by van Valen (1973), who borrowed the idea from a science fiction writer, Isaac Asimov, who borrowed it from Lewis Carroll's *Through the Looking Glass* (1865).

> 'Well, in our country,' said Alice, still panting a little, 'you'd generally get to somewhere else—if you run very fast for a long time, as we've been doing.'
>
> 'A slow sort of country!' said the Queen. 'Now, here, you see, it takes all the running you can do, to keep in the same place. If you want to get somewhere else, you must run at least twice as fast as that!'

The Red Queen hypothesis posits that organisms are in continuing ecological interaction with other organisms as they look for an evolutionary edge that will take them forward. An excellent example is the proliferation of bony armour and exoskeletons among the Cambrian animals, half a billion years ago. As the amount of carbon dioxide in the atmosphere increased, it became available for incorporation into protective armour as calcium carbonate. Potential prey became more heavily armoured; predators responded by developing more powerful jaws. Suddenly (in geological terms) a very rich fauna composed of these armorial novelties appears in the Devonian fossil record. These fossils are remarkable for the increasingly heavy armour on the one side – as in the thick carapaces of bottom-dwelling fish of the era – and the increasingly enlarging teeth and jaws – such as those of eurypterids or sea scorpions – on the other. But remember: this is not the whole story. Armoured forms are much more likely to leave a fossil record than the no doubt many equally successful soft-bodied groups. Hiding in a small hole in a rock is, for example, a very successful strategy against a large animal with an inflexible exoskeleton. This caveat, the imperfection of the fossil record, must always be borne in mind. It also applies particularly strongly to the human fossil history. Bones can only be found in sites suitable for their preservation, so we have a skewed idea of the social and biological evolution that was so important in humans.

Internal parasites are in an intimate and intense relationship with their hosts. Over generations, the parasite population exerts selection pressure on the host population, the host responds and, in turn, this exerts selection pressure on the parasite. This is not like a rally in tennis; rather, it is more like a protracted tug of war between two evenly matched teams, until a resolution is achieved. It can confer a surprisingly long-term stability on the host–parasite relationship. So much so that important evidence supporting continental drift was found among host–parasite relationships. For example, ratites make up a related group of large flightless birds. The South American rhea and tinamou, the African ostrich and the Australian emu all have parasites that are themselves closely related. The only reasonable explanation is that their ancestors were once cohabitants of the great continent of Gondwana that broke up and drifted apart.

The question that it raises is whether this family of parasites, having existed for at least 200 million years in the same family of hosts, is ever likely to develop mutually favourable symbioses. Clearly, they do not seem to harm the host species much, as host and parasite are still with us. It all depends on the nature of the relationship. The answer is no if, say, the offspring of a parasite depend on scavengers to eat the carcass of their dead host to provide them with a new home. Most parasites, however, have a vested interest in not killing their hosts for as long as possible.

Other parasite–host relationships long ago achieved equilibrium, to the stage where they cause the host no detectable harm at all. One example of a human host–parasite relationship demonstrates this point well. A study of the distribution of the parasitic protozoon, *Sarcocystis,* in Paris, France, found that more than 90 per cent of those tested had been infected by this protozoan parasite at some time and probably still were. It is thought that the Parisians owe this to their fondness for steak tartare, thinly sliced raw beef. In healthy people, the parasite remains dormant; only in cases of immunodeficiency does it show symptoms. So perhaps what we call a mutually beneficial symbiosis is the end point of long mutual struggle.

If so, it is a struggle that has succeeded over and over again. If a parasite – or anything – causes behavioural changes in an animal, they can, in turn, initiate evolutionary change by bringing that animal into contact with aspects of its environment it never before encountered.

Offspring can also vary by 'remembering' at a genetic level some of the events that occurred in the lives of their parents after their genetic cards had been shuffled and dealt and they have begun their own development. This statement would have been heresy until the last decade of the twentieth century and one would have been immolated in the righteous fires lit by the Darwinists of that time for uttering it. It is, of course, a form of Lamarckism, the 'inheritance of acquired characters', of characteristics acquired by an organism responding to pressures from the environment during its lifetime. Rechristened *epigenesis,* it has shed much of the opprobrium once heaped upon poor old Lamarck's head.

The justification for all the fury levelled at Lamarck was a concept called the Weismann barrier, named for the famous early evolutionary thinker, August Weismann. Weismann argued that reproductive cells were separated from body cells by physical and physiological barriers that could not be crossed. It followed, therefore, that events that occurred outside the reproductive cells could not contribute to inheritance. It sat at the heart of early twentieth century evolutionary orthodoxy. Even in 1942, however, Julian Huxley had qualms about the Weismann barrier and, in his great book that ushered in neo-Darwinism, he wrote 'the distinction between soma and germ plasm is not always so sharp as Weismann supposed'.

'Weismanism' is a fine demonstration of academic insularity, for even if the Weisman barrier were present it could only be in higher animals. Weismann had trained as a medical doctor and his anthropocentric prejudices were showing. Botanists were having none of the Weissman barrier. The plant kingdom represents more than 90 per cent of evolving life and botanists knew that, in plants, reproductive cells developed from somatic or body cells. The man who finally demolished the Weismann barrier and opened this can of zoological worms was Steele (1981), another paradigm shifter who suffered opprobrium for his view of the world. He showed that, in mice, changes in the parent's immune system may be manifested in the immune system of the offspring, 'pre-adapting' them in the event that they meet the phenomenon that brought about the change in their parents. What's good for the parents is good for the offspring and the likelihood of the offspring encountering the same challenges as the parents is high if they remain in the same environment.

In the last paragraph, the word 'pre-adaptation' is used advisedly. It is, however, a word that has spent most of its life being misused. It implies prescience in evolution, suggesting that somehow characteristics of an organism are already adapted to functions that they will perform in some future evolutionary scenario. This is obvious nonsense. Not every part of an organism is perfectly adapted to its environment. Arthritic back pain is the price humans pay for walking upright; we are certainly not perfectly adapted to living an upright life, in any sense of the word. The vertebrate spine works best in fish. Its effect is to prevent them concertina-ing due to water resistance as they swim forwards. It certainly was not was originally designed for vertical use.

Evolution, however, makes do with whatever it has handy. The evolutionary process is always making approximations, and those approximations depend very much on what has gone before and that in turn shapes the evolutionary future. We carry around with us many evolutionary relics of our ancestry. Sometimes, one of these relics is coopted by the evolutionary process and is modified into a useful adaptation. When that happens, we may say that the organism is 'pre-adapted' to its new condition as long as we remember 'pre-adaptation' does not imply that evolution had already earmarked it for future use!

Thus, pectoral fins originally allowed fish to plane up or down in the water and were there, ready to be used, for propping up the front end of a fish when it moved to colonise land. Later, pectoral fins became the forelegs of primitive amphibians and reptiles, the flippers of marine plesiosaurs, seals and dolphins, the wings of pterodactyls, bats and birds and the arms of humans. Fingers can be traced back to fin rays. These are all called homologous structures, having the same developmental origin. A crab leg, however, is an *analogous* structure to a mammalian leg; of similar function but derived from something else entirely. It is also important to remember that the visible parts of an homologous structure also have required simultaneous changes in internal structures, such as musculature and the internal architecture of the brain.

One of the great life changers is catastrophe. Catastrophism has been in and out of fashion since Noah's flood was used to explain the fossil record. This biblical idea was simply that every so often Jehovah got fed up with his creation, wiped the slate clean and started again. The alternative is gradualism. Today, as is now so often the case, instead of mutually exclusive gradualism or catastrophism, we can place them together in an interactive system without bothering Jehovah.

In 1972, Eldredge and Gould wrote an important paper called 'Punctuated Equilibria'. They argued that the *absence* of palaeontological evidence for intermediate forms in evolution was, in fact, itself evidence. They maintained that we had to believe the palaeontological record of evolution, because it was the only record we had. If it showed, in so many cases, long periods of gradual change interspersed with short, frenetic periods of diversification, then that was probably what happened.

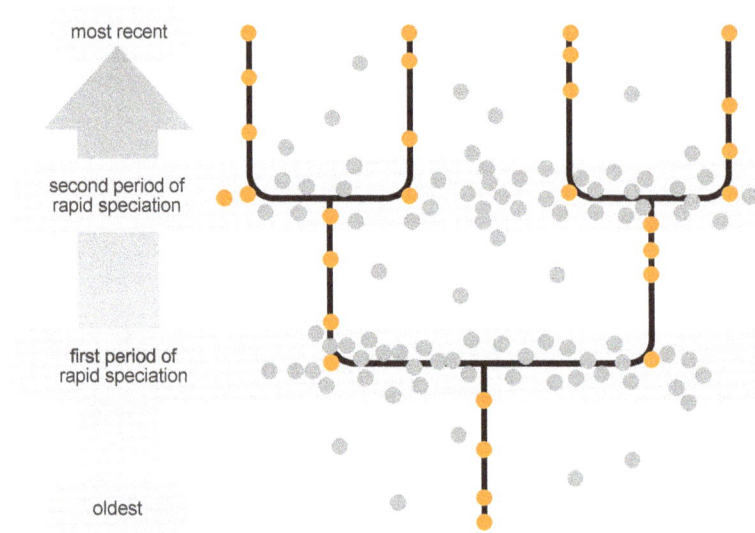

Figure 10. The punctuated equilibrium of a hypothetical fossil bed.

Figure 10 is the diagram of a hypothetical fossil bed, going from the oldest at the bottom to the most recent at the top. It shows punctuated equilibrium, layers where there are few fossils, indicated by circles, alternating with layers where there are many. Orange circles indicate the relatively few that have been 'found', the others are unknown and may never be known. A very neat tree showing possible relationships has been drawn through the orange circles, but it is not necessarily the right tree. Many others could be drawn.

'Punctuated Equilibria' disturbed two more heretical dogs in their uneasy slumber – first, the idea that evolution proceeded by a connected series of leaps and bounds. This was called *saltation* or jumping evolution. Second, it tended to resurrect the biblical teachings of *catastrophism*. In fact, the agitated response to this idea proved to be a storm in a teacup, and once again the sensible compromise prevailed. The evolutionary record is real, and various fossil lineages did indeed extend over long periods of slow change (stasis) interspersed with short frenetic periods of evolution into uncolonised evolutionary niches. At that point, different varieties of the same organism rapidly established themselves in a range of different environments that had now become available to them.

'Adaptive radiation' is the name given to this process by which the pioneers of every successful species become adapted to new circumstances. Many are called; few are chosen. The majority die but the survivors comprise the group on which natural selection acts to generate novelty.

Figure 11 is a fanciful representation of the Galapagos Islands, and bears no resemblance to their actual geography! They are the home of a group of small birds known as Darwin's finches because, as he was curating his bird collection, he was struck by the diversity of their beaks, providing him with yet another insight into the evolutionary process.

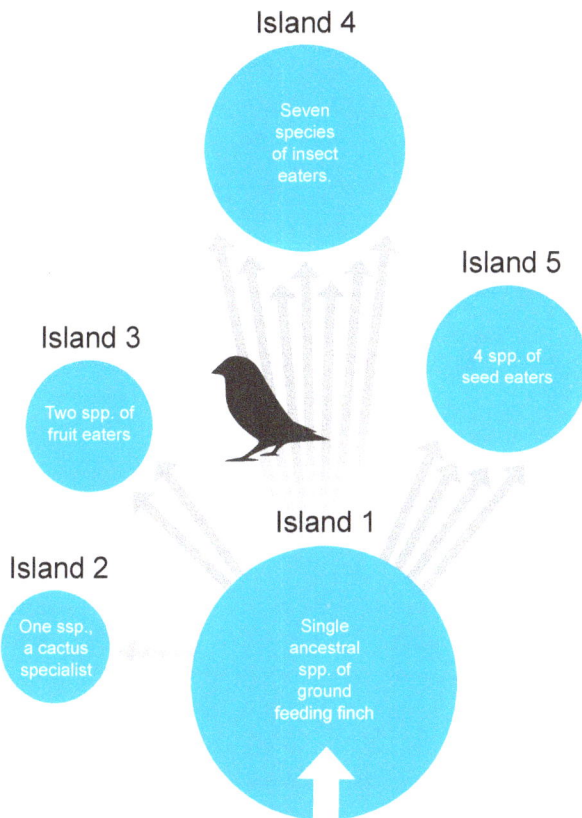

Figure 11. The spread of Darwin's finches.

On Island 1, the ancestral species is a ground-feeding finch. As its population grew, finches began to fly to other islands, or were caught up in storms and deposited there, or they rafted there on floating detritus. This must have been quite a common occurrence but not every finch survived the trip. For them to set up a colony, at least one of each sex was required. The finches managed to get to Island 4 a number of times and diversified into seven species. Only one successful colonising event occurred on Island 2, with its single cactus specialist.

The upshot was that after an unknown number of generations, each island had its own unique populations of finches, distinguishable by the shape and size of their bills.

The new environmental conditions that permit adaptive radiation are brought about by various causes, one of which is indeed catastrophe. There is no escaping the effects of random events such as asteroid strikes over long periods of time. Extinction is a great stimulator of novelty by creating evolutionary space. The first great extinction occurred with the release of toxic oxygen into the atmosphere by ancient photosynthesisers, an event that put anaerobic microbes under intense selection pressure and killed off many of their species.

From counting known fossils, it appears that there have been at least five other *great* extinctions. Fossil counting is, of course, an imperfect measure, because not everything gets fossilised, especially organisms with no hard parts, and only a fraction of the possible total of fossils has been collected. It is, however, the best that can be done.

Figure 12 is a plan of the geological ages from the Cambrian, 500 million years ago, to the present. Figure 13 gives a rough suggestion of dominant life forms at various times. At the end of the Cambrian, 450 million years ago, the fossil record suggests that two events killed off 65 per cent of all species. After another 80 million years of recovery, 70 per cent of all species were lost in the Devonian. The largest event of all, at the juncture of the Permian and the Triassic, 250 million years ago, did away with 90 per cent of all species and even the colonisation of land by Amphibia was in jeopardy. The Triassic/Jurassic extinction followed 50 million years later, and 75 per cent of all species vanished. There were already evolutionary experiments going on, such as mammal-like reptiles and feathered dinosaurs, and the first true mammals and the birds took full advantage. The groundwork was laid down here for complex immune systems to fight bacteria and enable temperature regulation at around 38–40°C. This is the temperature at which birds and mammals function best, but sadly it is also the one at which bacteria grow best.

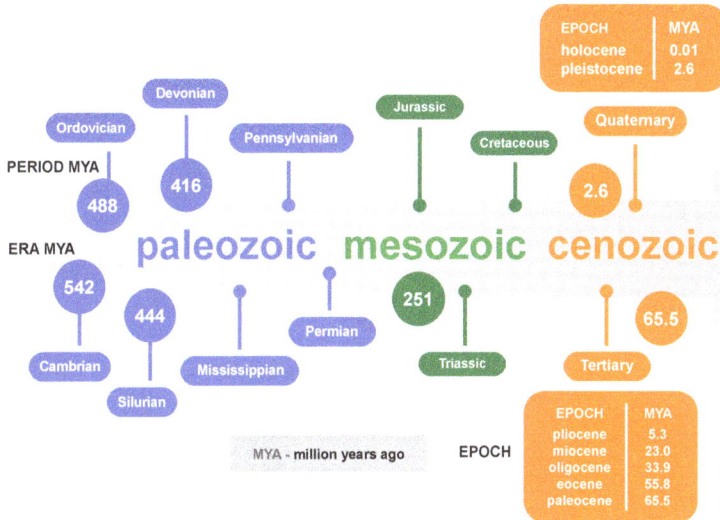

Figure 12. Geological ages from the Cambrian period, 500 million years ago (MYA), to the present.

The most notable extinction was caused 66 million years ago by the great asteroid strike on Earth, made famous by Luis and Walter Alvarez and colleagues (1980). It caused the extinction of 75 per cent of all known species. This discovery alerted everyone to how precarious was life on Earth. It is a chilling thought that another big asteroid is due in a few million years or so, unless we can achieve the technology necessary to divert its course.

Add to all this mayhem the so-called lesser extinctions – about 20 of them, not including the one currently being precipitated by humanity – and the odds against any one genetic line making it through from the origin of cells to the present day are enormous.

But hey! Look around. We all did it, from the meanest bacillus to the blue whale! Gaia is nothing if not resilient. She is, after all, a 'tough bitch' as Margulis colourfully described her, in a book of essays edited by Brockman (1996).

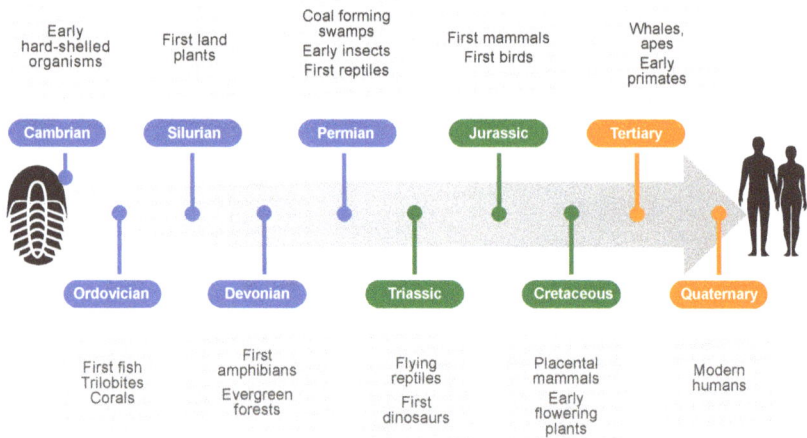

Figure 13. Dominant life forms across the geological ages.

Another great source of change for living organisms lies in the possibility that the rates of development at various times in their life cycles can vary. Every organism must be adapted to its environment at every stage of its life history. A common example is the cabbage white caterpillar that lives in an entirely different world from the butterfly into which it metamorphoses. In Lepidoptera, the caterpillar is the form that does the heavy feeding and the butterfly form does the breeding. Both caterpillar and butterfly must 'be fit for purpose', adapted to the life they lead. In fact, in these insects, the caterpillar spends most of its life feeding in that form, turning into the relatively short-lived winged form only to seek a mate and breed.

Even in organisms that do not experience the remarkable shape change of the butterfly, development does not always occur at the same rate throughout its growth phase. Sometimes it slows, at other time it accelerates, the changes in each phase having consequences for the form of the adult. If the organism passes through markedly different juvenile forms on the way to adulthood – the larval form, say, in animals or the production of juvenile foliage in plants such as the eucalypts – these different developmental forms must be as well adapted to their environments as the adult forms. At the beginning of the last century, Edith Johnson (1926) showed that the two sorts of leaves in *Eucalyptus globulus* worked in different ways. The juvenile leaf faces the sun but is much more likely to be shaded by other shrubs than the adult one. It has a thicker layer of wax to inhibit water loss by transpiration and also is less easy to wet, reducing the chances of fungal attack. The adult leaf, on the other hand, tends to be edge-on to the sun, has no wax but it does have a thicker cuticle and a greater rate of water transfer. There are other differences but, unsurprisingly, it seems that the juvenile leaf is adapted to life in the understorey of the forest, while the adult foliage is adapted for life in the canopy.

A forest tree thus grows through two environments, first as a bush and then a tree, and exploits each while staying in the same place. Marine animals, like sea squirts and barnacles, that stay in the same place as adults, do not usually have that luxury. The strategy they adopt is to insert a larval stage to into their development. There are many examples of larvae – for example those of crabs, mosquitoes and frogs – with lives different from their adults. These larvae must be capable of surviving their free-living phases – that is, be adapted to their environment. For animals where the adults are cemented into position, like the corals and barnacles, the imperative for the larva is to survive long enough to find a safe living space, where they can settle and develop into reproductive adults.

It is possible to think of the human foetus in this way. Humans do not have larvae but given the immature and unprotected state of small children, they might as well have. Organic evolution, however, has been complemented by social evolution. Adults provide their children with food, clothes, shelter and a collaborating society that allows humans to adjust to changes in the environment without going the direct genetic route. The sequence of agricultural revolutions that created stable environments is part of this and comprises a major evolutionary advance for the human species.

Varying the relative lengths of a life history so that an organism becomes adapted to two or more environments in its lifetime is thus a common evolutionary trick. Larval mayflies spend months living under stones in stream beds, feeding, and only a day or two as winged insects whose function is reproduction. One species of cicada (*Magicada*) spends 17 years underground with its mouthparts stuck into tree roots, feeding. The adults emerge to reproduce and live about four months. Another animal, the liver fluke (*Fasciola hepatica*), has three quite distinct forms to cope with three different environments during its life. A close relative has four!

Every stage of an animal's – or plant's – life history is therefore tested by natural selection and that includes its *duration*. It may be either lengthened or shortened. One way is paedomorphosis, meaning literally 'child form', the retention of an early developmental stage into adulthood, when the animal becomes reproductive. It is an idea that was particularly well developed by marine biologist NJ Berrill (1955), who suggested that it was responsible for the origin of vertebrates. This was because the free-swimming larvae of sea squirts had a number of vertebrate-like characteristics. Berrill suggested that some larvae long ago failed to turn into adult sea squirts and became sexually mature, thus affording access to a whole new evolutionary space now occupied by vertebrates. The present view is that he was nearly right, but it is more likely the vertebrates are derived from a sister group.

Neoteny is said to occur when the physiological or bodily development of an organism, usually an animal, is slowed or delayed. Ultimately neoteny results in the retention of juvenile physical characteristics in the mature adult form of a species. The classic modern example is the axolotl, which looks like a giant larval salamander when it is sexually mature, while having retained its larval traits.

Another example of neoteny is our own species. Structural human neoteny is a respectably old concept dating back to the 1920s. There are a number of human features that are characteristic of foetal apes, and it is difficult not to conclude that humans are neotenic apes. The process of neoteny seems to have been led by the evolutionary advantages afforded by a big brain, while constrained within a simian morphology. We discuss this at greater length in Chapter 11.

In humans, the juvenile growth stages, childlike characteristics, are maintained into adulthood. They do not take a physically different form except in relative proportions as in, say, head circumference and length of arm. It is possible, however, that brain development is discontinuous. Piaget made a strong, although contended, case for the following stages of human development:

1. Age one to three years, in which the individual believes, and acts, as if they are the centre of the universe.

2. At about seven years, they are able to recognise that the world can change, and they can recognise the change.

3. For most, it is 12 years or more before the child can think or speak of abstractions, such as peace, humility, ageing and other concepts.

The development stages are cumulative, unlike the butterfly that dispenses with the earlier form!

Recently, the concept that each organism is not an individual but a collaboration of many different sets of genes, has been incorporated into a hologenome theory of evolution. It is hotly debated (Bordenstein and Theis 2015; Moran and Sloan 2015). It proposes that natural selection does not occur at the level of an individual organism but at the level of what has been called a 'performance unit'. A performance unit comprises a primary biont – such as a cow or a human – and its population of symbiotic microbes. That is to say, in my case, the primary biont is what I think of as 'me' and the secondary ones are all the hangers-on on my skin and in my intestine. In short, what I think of as 'myself' is a community, a holobiont.

This is an important change of perspective. The unit of evolution becomes a community, a set of cooperating and interdependent individuals – or genomes. This is indeed the triumph of the world of ancient cells, if you count the modern cells of a human being as ancient cell constructs. The population structure of microbes in the holobiont, however, is neither fixed nor necessarily totally predictable, and so the hologenome community depends on circumstance.

The acquisition of, and providing homes for, various microbes can give animals and plants access to new resources – the ability to consume a new kind of food or to survive in a different environment. This is well illustrated by the termites and the ruminant mammals, whose intestinal denizens allow them to digest cellulose. When this happened the first time, a whole new world based on cellulose as a food source was opened up. Plants also acquire symbionts and fail to thrive without them; for example, the fungal root nodules that fix nitrogen in legumes and many other beneficial plant–microbial interactions.

Future technology will always looks like magic from the viewpoint of someone 50 years in the past. If you had told someone in the 1960s that you could tell who and what had been swimming in a river merely by examining a few litres of its water, they would have thought you either mad or a magician. Life is, however, deciduous, always shedding little bits of its DNA everywhere. Today, the magic can be done by collecting the minute fragments of DNA the water contains, replicating them over and over again by a simple, repetitive chemical process and asking a computer to identify them by comparing them with its database of known organisms. This process is now a valuable ecological tool.

There are other 'rivers', such as the one consisting of food breakdown products and liquids that flows through your intestine. Easily, if embarrassingly, sampled, the DNA there provides evidence of the many hundreds, maybe many thousands, of different microbes that contribute to the health of the holobiont that you are pleased to call 'I'. And many of them have been unknown to science.

One final important point: holobionts, complete with fellow travellers, *do not change the rules* of evolutionary biology. Although these concepts redefine what constitutes an individual animal or plant, they do not require a fundamental rewriting of Darwin's theory of evolution by natural selection. Rather, they increase the significance of considering both parts and wholes as sets of fruitful relationships.

5

THE PAST IS A FOREIGN COUNTRY

In which we argue that, while the historical reductionist approach to science has had its triumphs, it has been shown to be inadequate in exploring complex issues.

Before Darwin and the great synthesis that was *On the Origin of Species,* it had been common to reduce any significant problem to its individual strands for separate study. In this chapter we explore the rationale behind this reductionist response to evolution as it was played out in the nineteenth century, before the return to a wholistic understanding.

In the 1950s, approaches to knowledge were at a watershed, the point at which a reductionist past began to separate from a wholistic future. In no spirit of criticism, for it was a creature of its time, this event is nowhere better illustrated than by the Honours course in zoology at King's College London from 1955 to 1958. The structure of DNA – the ultimate reductionist triumph – had been published only two years earlier in *Nature* but academic courses had not even begun to come to terms with the implications of this astounding discovery.

So, undergraduate zoologists in those years still assiduously took animals to bits and examined the bits. They carried out simple physiological experiments on some of the bits, such as frog muscle, or tortured amoebae in Petri dishes. Student botanists were doing the same sorts of things on the floor above with plants. It was very enjoyable, and quite Victorian, as some of the apparatus had been in constant use for 100 years.

There were, of course, occasional 'enrichment' lectures from savants from other universities, and students were taken out into the field to complement their laboratory studies with a bit of natural history. Fieldwork consisted of traipsing around the English countryside armed with tape measures, quadrats, pond nets, collecting bottles and pooters (also called aspirators; these last were simple glass devices with which you could suck up by mouth, without swallowing, anything for later study that was too small or too venomous to handle). It was good fun, but there was not much theoretical basis, beyond elementary statistics; it was strictly a matter of seeing what was there and counting it. Systematic zoology, slotting things into their separate taxonomic boxes, was the major objective. The leading metaphors that applied to evolution tended to be those of mathematical genetics and palaeontology and Kipling's *Just So Stories*. However, the sun was setting on this predominating method of inquiry.

At the end of their three years, students were rewarded with graded degrees. This put them in *their* boxes, where they were expected to stay for the rest of their lives. The 1955–58 group only remained in their boxes until the 1960s, when the desperate need for the multidisciplinary approach was becoming clear.

At the time there was some excuse for continuing with reductionism. The science of modern genetics – now called genomics – had made its first steps as the functions of the nucleic acids were worked out and the first cautious identifications were made of the 'letters' in the DNA code. Elsewhere, the *Fundamentals of Ecology* by Eugene Odum and *Distribution and Abundance of Animals* by Andrewartha and Birch had just been published in 1953 and 1954, respectively. Rapidly developing postwar technology was beginning to make possible the elementary steps of a wholistic science that permitted the study of complex systems. The wholism–reductionism dichotomy was still often evident in conflicting interpretations of experimental findings and in the setting of competing priorities for future research.

It is essential to understand that there is nothing intrinsically evil about reductionism. In fact, humans are probably born both reductionist and wholist. It is perfectly natural for a child to break something to see what it is made of or take something to pieces to see how it works. At the same time, they see the world as a whole, centred on themselves. Most people move on from that position as they acquire the dominant knowledge of their own era. It is therefore impossible to assign the philosophy of reductionism to a particular place or time of origin. A useful historical starting point for discussion is with a fourteenth-century monk, William of Occam (or Ockham) and his Law of Parsimony (Occam's Razor). It states that, in reasoned argument you must not increase the unknowns. In other words, work only with what is known. Two centuries later, René Descartes, one of the key figures in the science revolution of the Renaissance, used this rule to promulgate a very mechanistic view of biology. For example, he believed that animals were merely automatons made of meat and rejected any view that posited purpose in their existence.

Reductionism has proved to be a useful tool in science, which progresses by accretion of knowledge, so that arguments improve as more of the unknowns become known. The contrasting view to reductionism is wholism: the idea that things can have properties as a whole (emergent properties) that cannot be understood from a simple knowledge of their individual parts. Even a quick look at evolutionary history suggests that the evolutionary process is a series of emergent phenomena.

Emergent properties are generally the properties of complex systems, whose complexity is the consequence of many simple, reiterated, recursive interactions. Every major evolutionary event has led to consequences that a contemporary observer, from Mars, say, could not have predicted. The fables in Chapter 3 emphasise this point. The whole of the biosphere, including human social systems, is an emergent consequence of the appearance of the first cell.

To do them justice, many reductionists understood this phenomenon, but chose to ignore it in the process of studying what was possible, given the state of science at the time. Up to the middle of the twentieth century, the study of biological and human systems as a 'whole' was difficult, unreliable and time-consuming. It created an unfortunate but pragmatic situation, where things were studied more because they *could* be studied, rather than because they necessarily *should* be.

The reductionist approach to humans is epitomised by the study of anatomy. There, the process of dissection, of separation into parts, made studying the whole living organism impossible. The physiologists followed suit and not until the end of the nineteenth century did scientists shake off the pseudosciences such as phrenology (determining character from the shape of the skull) and phlebotomy (treatment of disease by bleeding) and begin the attempt to study *Homo sapiens* as an integrated whole. Ironically, the medical profession has achieved great success by developing blood tests for known disorders, a process arguably wholist (identifying the disease) and reductionist (creating the blood test and prescribing a remedy) at the same time.

The advent of genomics, the study of the highly variable genetic kit owned by all organisms (see, for example, Lesk 2017), brought about another wave of scientific reductionism. Readily available 'cookbooks' gave the simple and detailed instructions for gene sequencing and manipulation. Graduate students were exploited as intelligent workhorses to do the menial task of gene and protein sequencing. The cynical slogans 'one polypeptide chain, one PhD' and later, as techniques evolved, 'one gene, one PhD' were current around the turn of the twentieth century!

The past is indeed a foreign country and, as LP Hartley remarked, they certainly did things differently there. The sense of the connectedness of things was lost during the Enlightenment when the scientific method of destructive analysis became de rigeur. Philosophers believed that understanding came from dissection, and much understanding did come. By unweaving the network of knowledge into its component threads, the philosophers of the time were so intoxicated by their so-called objective discoveries that they lost sight of the whole. Once the metaphor of individual creation by a higher being was recognised as untenable, it was replaced by another dangerous metaphor, evolution. Darwin's plea for his tangled bank was forgotten and life on Earth came to be seen as a battlefield of relentless competition among all forms of life and the story one of a struggle for existence.

This pessimistic view of life gave rise to the popular evolutionary catchphrase, 'the survival of the fittest'. It was not of Darwin's invention although he did use it from time to time. It was a convenient dictum, erroneously interpreted and applied by many to the human condition. Poor people were considered, *because* of their poverty, to be less fit – but less fit for what? In Victorian times, competition had been held as the way to social and financial success. It had been invoked in empire-building, business, sport, in the establishment of class structure and in people's identity. It had thus been easy to distort Darwin's message further and portray competition as the natural and most desirable condition towards which human beings can aspire.

The authorship of this unfortunate phrase, 'the survival of the fittest', is ascribed to Herbert Spencer, a great if somewhat misguided supporter of Darwin. The most pernicious use of this catchy war cry occurred when

Francis Galton adopted it as a justification for his sociological theories. Out of these came eugenics, and while Galton took a positive view by encouraging 'above average individuals' to marry and have children, it left open the question of who decided who were these 'above average' people who were to be the parents of the brave new world – and left unanswered what to do with the 'less than average'. The answer to the question of who were above average usually proved to be 'us', the Establishment. Claiming that this was 'scientific' implied, erroneously, that it was objective and the decision could be made without sympathetic or ethical considerations.

The unstated corollary, that the below average – which came to mean 'not human like us' – should be prevented from breeding, led to it being used to justify the horrific excesses of the Second World War and to subsequent cases of wholesale murder euphemised as 'ethnic cleansing'. So, at the end of the nineteenth century, the new biologically based hypothesis of the primacy of social competition in evolution became very much the vogue. It was followed in the early twentieth century by the belief in 'the perfectibility of man' based on an optimistic enthusiasm for human social engineering, eugenics. In Nazi Germany, this heady brew was laced with a large dollop of Wagnerian mysticism. The quest for the true Aryan – blond, blue-eyed, muscular and fit – led those not conforming to this ideal straight to the concentration camps. The Russian cartoonist Boris Efimov lampooned this vile creed in 1941 with a picture of the fascist big three and the caption 'A true Aryan should be tall like Goebbels, muscular and slim like Goering and blond like Hitler'. Now, in the twenty-first century, social Darwinism can be recognised for what it is, a distortion of Darwin's theory into a justification for 'unnatural' selection.

The nineteenth century was a time of great change. As noted above, natural philosophers had been flirting with the idea of evolution for some time but no-one had been able to suggest a satisfactory mechanism for it. The Christian Establishment was still more or less intact, although Thomas Robert Malthus, himself a churchman, had, in 1798, set the bomb under it that would eventually undermine the fundamental base and bring the edifice down. In his *Essay on the Principle of Population*, Malthus had grasped the importance of exponential growth, in which the size of the population progressively doubles, until all resources are used up (Figure 14). He wrote that 'the power of population is indefinitely greater than the power in the earth to produce subsistence for man'. This proposition remains especially relevant today.

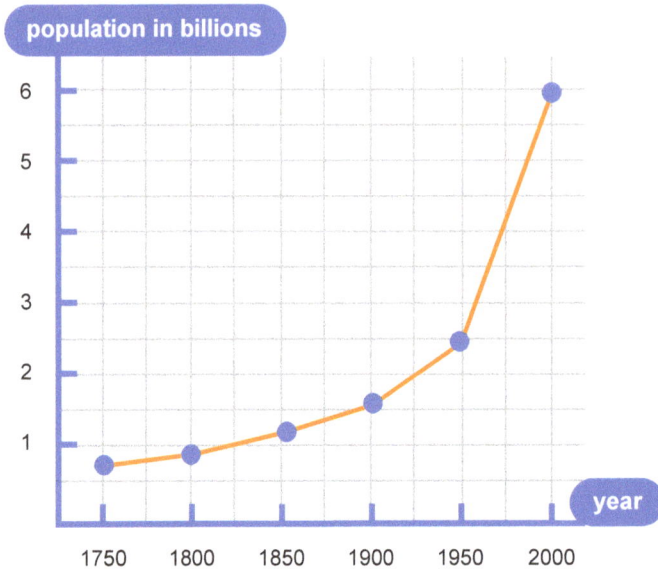

Figure 14. Human population growth since 1750.

Malthus's seminal paper inspired Darwin's search for a mechanism of evolution, although it is focused on the selective pressure created by the growth of every population of organisms, a 'single-factor' explanation of a complex system of relationships. It suited the thinking in Victorian times and still resonates. The proliferation of the poor was used to justify the belief in a hierarchy of rank and privilege underwritten by wealth, by God's will.

There is a common view of life that is unnecessarily pessimistic. Consider the following words from *Last of the Summer Wine,* a low-key but highly successful BBC TV comedy program. Three ageing Yorkshiremen are leaning on a five-barred gate, overlooking a beautiful Yorkshire dale in early summer and musing on the meaning of life. The character of Norman Clegg, a self-made philosopher of penetrating observation, having commented on the beauty goes on remark that 'faintly on the breeze, you begin to sense the million munching teeth of tiny things scoffing even tinier things'.

In this line, spoken with delicate nuance by that fine actor, Peter Sallis, Clegg is expressing a commonly held view, that 'ecology is a terrible book in a beautiful cover', that 'red in tooth and claw' is a foundation for the beauty of nature. And, indeed, it would be depressing, if it were true. In fact, it is a very animal-based view and takes no account of the overwhelming biomass of green plants and fungi, and the myriad instances of cooperation between organisms – including humans. Even if cooperation between members of the same species is ruled out (which it cannot be), there are so many other examples of inter-species cooperation to give the lie to this pessimistic view. Even at the turn of the century, Kropotkin (1902) could write 'don't compete! – competition is always injurious to the species, and you have plenty of resources to avoid it!'

In the middle of the Victorian century, it was unfortunate for the popular conception of evolution that one of the great Victorian writers, Alfred, later Lord, Tennyson, was also grappling with these ideas and got in first, before Darwin. It is a pity, because his imagery, given all the impact of the art of a great poet, stays with us. Tennyson was, in the current vernacular, in a bad place when he wrote a threnody for his close friend, Arthur Henry Hallam, who had died unexpectedly. *In Memoriam A.H.H*, completed and published a decade before *On the Origin of Species,* expresses Tennyson's failure to reconcile an uncaring Nature with the Christian God of Love. In the following verses, she, Nature, is a metaphor for the extinction of previous life forms:

> 'So careful of the type?' but no.
> From scarped cliff and quarried stone
> She cries, 'A thousand types are gone:
> I care for nothing, all shall go.

Those who, like Hallam, clung to a belief that nature was love, were conflicted by the thinking of the time:

> Who trusted God was love indeed
> And love Creation's final law –
> Tho' Nature, red in tooth and claw
> With ravine, shriek'd against his creed –

Hugh Miller, the greatest palaeontologist of the time, shared Tennyson's internal torment. Miller was the stonemason author of *Old Red Sandstone.* His was the seminal and encyclopaedic work, first published in 1851 (*Everyman* edition 1919), on the Devonian fossils from 'scarped cliff and quarried stone', which he compiled to illustrate the glory of God. Already depressed, it is said that when he learned that his own work was one of the main planks in the evolutionary theory being developed by Darwin, he shot himself.

Like social Darwinism, Tennyson's trope, 'nature, red in tooth and claw', outlived its context and reverberated down through the twentieth century. It is a strongly zoocentric and negative point of view and ignores the great bulk of evolutionary creation in which life forms interact to mutual advantage. In the plant kingdom, for example, death is not the necessary end product of the depredations of a nibbling caterpillar. In the animal kingdom, large carnivores live at peace with one another and often share the same territory by occupying different ecological niches; lions and leopards coexist because they hunt at different times of the day for different prey.

Darwinism has always been cursed with people who try to encapsulate in a few words the important concept of evolution. Misunderstanding the evolutionary relationship of humans to apes, Bishop Wilberforce poked fun at Thomas Huxley by asking him whether it was through his grandfather or grandmother that he claimed his descent from a monkey. Huxley's response, probably apocryphal, has also echoed down the years: 'I would rather have a monkey as an uncle than a man who uses his intellect to obfuscate an argument'. As we shall find in later chapters, the idea that humans are descended from monkeys is so nearly right, while being hopelessly wrong, that it survives today.

Herbert Spencer's 'survival of the fittest' also became a catchcry – probably every English-speaking secondary school child can trot it out if asked about evolution – but in itself, it is meaningless and adds nothing to the debate as it is self-referential and circular. (Who are the fittest? The ones that survive. Who are the survivors? The ones that are the fittest. And so on.)

The search for the 'missing link' between monkey and man had become so imbued with evolutionary politics by the turn of the century that someone went out of his way in 1908 to provide one in the shape of the artfully constructed fossils that comprise 'Piltdown Man'. Although the comparative anatomists had their suspicions, Piltdown Man was not finally exploded until new methods of dating were available in the 1950s. The skull was found to be only 500 years old, that of an evolutionary modern man, the lower jaw came from an orangutan and the filed down teeth from a chimpanzee.

A better term is the 'struggle for existence', but even that is not very good. Struggle can be interpreted in many ways and too often is assumed to mean the struggle between members of the same species for resources. But more often the struggle is against the environment: against wind, rain, drought, poor soils, fire, frost, heatwave, earthquake, tsunami, asteroid strikes, disease and random events. A random event is anything we have not thought of in this list, such as falling out of a tree! This was already, in 1905, apparent to Peter Kropotkin in his travels across Northern Europe. The attention of the Russian zoologist was caught by the numerous cooperative interactions he encountered among the birds and mammals in the harsh environments of the north. This so impressed him that he recorded his observations in a major essay entitled *Mutual Aid: A Factor of Evolution*. Unfortunately, this too was seized on to support political objectives, this time of anarchism, guaranteed to startle the horses of the Establishment.

As a general principle, natural selection as a foundation for the evolutionary story has stood the test of time. Evolutionary change through natural selection has been accepted as shaping our everyday physical and social reality, very much as proposed by Charles Darwin a century and a half ago. On the other hand, the story it told continued to generate highly imaginative ideas, built on incomplete evidence, and interpreted through the historical and cultural lenses of the day. For instance, in the nineteenth century, when the idea of a struggle for survival leapt to prominence (although not initiated by Darwin), it was immediately used to justify the power hierarchy and to impress on impoverished people the feeling that their condition was divinely ordained and all they could expect. It was a convenient reminder for the poor to invite them to sing, during their devotions in church, the verse of Mrs Cecil Alexander's 1848 hymn 'All Things Bright and Beautiful':

> The rich man in his castle,
> The poor man at his gate,
> God made them high and lowly
> And ordered their estate.

The authors remember singing this verse with gusto and enjoyment in their primary schools; happily it is rarely sung today. The hymn was the subject of a sardonic and satirical parody, 'All Things Dull and Ugly', by Monty Python.

The Victorians, presiding over a vast empire held in check by military power, were prone to see everything in terms of the metaphor of conflict, a vision reflected in their hymns – 'Fight the Good Fight', 'Onward Christian Soldiers', 'Soldiers of Christ Arise and Put your Armour on', to name a few. Charles Darwin himself had a much softer view of nature than most, but he could still write at the end of *On the Origin of Species* that:

> from the war of nature, from famine and death, the most exalted object which we are capable of conceiving, namely, the production of the higher animals, directly follows. There is grandeur in this view of life …

Note that this too is animal-centred, approaching the world from the perspective of the animal kingdom, without acknowledging the contribution of plants, or the social constructions of *Homo sapiens*. It does, however, acknowledge and warn that any approach to the concept of evolution by natural selection is necessarily anthropocentric, that is, human-centred, being a product of human thought. And, as we have seen, its interpretation is always through the social understanding of the times. In the tradition of reductionist science trying to eradicate subjectivity, it has meant that evolution has been interpreted as a given biophysical condition, thus ignoring the contribution of humans, both as a major influence on, and interpreter of, the evolutionary process.

In 1900, the rediscovery of Mendel's important study of inheritance in peas at last provided a mechanism by which natural selection could operate. In sexual reproduction, there is a re-sorting of each individual's genetic program at each generation. Three independent scientists rediscovered and republished Mendel's findings in their own countries within a few months. In England, Hugo de Vries was given the credit for the disinterment. Mendel's contribution was the discovery of 'particulate' inheritance, of the gene. Characteristics specified by these particles were passed to the next generation intact, not in bits to be blended, as had previously been supposed. A good account of the impact of Mendel on biological thought is to be found in Edelson (1999).

Another 'heresy' put to the sword in this period was so-called Lamarckism. Summarised as the inheritance of characteristics acquired in a lifetime, Darwin viewed it quite kindly. With, however, the important discovery by Mendel of the inheritable effects of genes, scientists lined up in their traditional adversarial positions. It had to be either one or the other: Darwin and Mendel in one corner, Lamarck in the other. In a *cause celebre* after the Great War, in 1926 the biologist Paul Kammerer was accused of falsifying evidence to support the Lamarckist position on environmental influences on inheritance. Kammerer was probably correct in what he observed, but his Lamarckian interpretation at that time was considered to be heretical (Koestler 1971).

Today, a rather different version of the Lamarckian view is widely accepted and made respectable as 'epigenesis', from the Latin meaning on or around the genes. Epigenesis is the idea that environmental changes experienced during the lifetime of a plant or an animal may also affect its offspring (Francis 2011). As a concept it is not terribly controversial today. But in 1926, the battle lines were drawn. Few thought in terms of this *and* that.

'Darwinism *and* Lamarckism' is a phrase that was rarely encountered in a positive way in a single sentence in the last century. The more common 'Darwinism *or* Lamarckism' embodied the confrontational nature of science in those days – two sides, drawn up and ready to do battle, a battle that the Darwin supporters turned into a rout. It was a very asymmetric war. A great deal of the evidence for Darwinism was already available while Lamarckists were readily dismissed as cranks.

Not until the end of the twentieth century did modern scientific techniques precipitate a serious resurrection of interest in Lamarckism. Then, with a nod to the Frenchman, it was rechristened 'epigenesis' in *Lamarck's Signature* by Steele et al. (1998), who described a way by which information from non-reproductive cells can be inherited. The authors, self-proclaimed Darwinists, were not trying to replace Darwinism with Lamarckism – their aim was to modify modern evolutionary theory to include some Lamarckian phenomena. Such collaboration had not been permitted before.

We know today that no DNA comes into the world naked. In the case of bacteria, the two daughter cells resulting from a cell division partake evenly of all the pre-existing cellular paraphernalia. In the case of reproduction by gametes – human reproduction, say – the maternal egg possesses all the machinery for cell growth and division, needing only the stimulus and the DNA provided by the sperm cell, which inserts only its nucleus into the egg to begin development. All cellular membranes thus come from pre-existing membranes provided by the egg, needing only their capacity for self-assembly to equip the new organism.

With discovery of the genetic code in DNA by Watson and Crick in 1953, the reductionist approach was given another lease of life. Immense effort was put into deciphering the code, determining which triplet of bases coded for what amino acid, learning how coding took place, how the code was translated into a polypeptide chain and the roles played by the various RNAs. This phase of research was well underway by 1970, and molecular biologists began to manipulate genes to find out how they were regulated and then went on to sequence complete genomes. The first organism to have its genome 'read' was a free-living (as opposed to parasitic) nematode worm, *Caenorhabditis elegans*, in 1998, an achievement that won Sydney Brenner a Nobel Prize, reported by Check (2002).

The past reductionism that considered plant and animal inheritance solely as the consequence of sexual recombination supported by mutation and genetic drift is long gone. Evolution by natural selection can now be considered chemical (the expression of genes in the dance of RNA and associated protein molecules around the DNA), environmental (selection pressures generated by changes in the real world), social (the influence of interaction with organisms of our own and other species) and historic (epigenetic – changes to the genome brought about by life experiences) and more. The introduction of the three-letter word 'and' between items on that list creates an entirely different evolutionary landscape. The concept of evolutionary change as the wholistic outcome of multiple interacting factors is now widely accepted, and is providing the justification for rethinking Darwinian evolution in the twenty-first century.

As an example of reductionism as a success, consider biochemistry and cell physiology in the middle of last century. The reductionist approach was to try to take cells and tissues apart to isolate the subcellular components and make tissue extracts in an effort to determine their function. A common technique was to take the tissue of interest, such as rat liver, add a concentrated sugar solution and blend them together in a kitchen blender. Having broken down nearly all the structures in the cells of the liver, the result was placed in a tube in a centrifuge and spun at about 1,000 times the force of gravity. The pellet formed at the bottom of the tube was discarded and liquid was transferred to another tube and centrifuged once more, this time at much higher g-forces. A beige-coloured pellet was obtained that proved to be composed of millions of energy-producing organelles, called mitochondria, from the liver cells. Mitochondria thus prepared were very stable and, under the electron microscope, looked much the same as they had done when in the liver.

In the reaction flask, under relatively simple experimental conditions, they could be made to carry out their normal cellular function of energy production that previously they had performed within the cell.

Much earlier, in 1890, Altmann, using a light microscope, had noticed granules within the cytoplasm of cells that he thought might be bacteria. With the advent of electron microscopy, the bacteria-like granules turned out to be mitochondria. Altmann's observation was an inspired anticipation of Margulis's symbiotic theory of the origin of eukaryotic (modern) cells 80 years later. Her work, a return to the study of the whole, led to the recognition of the cooperative events that underpinned multicellular life throughout the living systems of the planet, of which more in the next chapter.

There is one discipline, that of human palaeontology, where, in its beginnings, the thread of evolution was invisible. The scarcity of human fossils at the time of *On the Origin of Species* led to much interest in the so-called 'missing link' between the anthropoid apes and humans. There are now many other examples of 'missing links' that are not, in fact, missing, in different evolutionary lines such as the living ginkgo tree that takes us back nearly 300 million years to the era of the cycads, or the fossil *Tiktaalik*, a so-called 'fishapod' that bridges the gap between fish and amphibians, or the fossil *Archaeopteryx* that relates dinosaurs to birds. The human 'missing link' was seized on by the religiously inclined. The fact that it was missing reinforced their entrenched view of the uniqueness and divine origin of humans, even though there are, in fact, many living species that might be considered to be intermediary between non-human primates and humans.

Today, Darwin's work stands as a bridge between the biophysical understanding of evolutionary change and the recognition of the role of human social evolution. In his book on the *Expression of Emotions* (1872), he accepts the human dimension as a major contribution to evolutionary change in its own right.

6

WE DO THINGS DIFFERENTLY NOW

In which the reductionist view of the world is overtaken by consideration of the system as a whole.

The various biological disciplines began to reconnect from about the middle of the twentieth century, after Julian Huxley published *Evolution: The Modern Synthesis*. The world had changed: most scientists accepted Darwinian evolution as a 'given' and humanity has, latterly, become aware that it is itself a species at risk from its own ability to wreak environmental destruction. From a wholistic perspective, the world had been recognised as a continually changing pattern of self-organising, cooperating systems involving all life. And, as Dobzhansky (1973) remarked, nothing in biology made sense except in the light of evolution.

The second half of the twentieth century therefore saw a serious challenge to the reductionist habit of breaking things into their pieces. By then public opinion was moving forward from the heritage of the scientific Enlightenment to adopt the idea of wholist interpretations of events. A social movement embodied these concepts: the 'flower people' of the 1960s, the hippies, proclaiming the ideal of universal belonging. Wholistic science followed. It recognised that, in a complex, fast-changing world, the important questions could not be answered by studying their parts alone. Food security, a stable atmosphere and the integrity of the oceans are only a few of the complex issues that urgently needed, and still need, to be dealt with as a whole.

You will by now have noticed that we have consistently used the traditional spelling 'wholistic', to describe those relationships which form a whole that is not only greater than, but also different from, the sum of the parts. The word dropped its 'w' as recently as 1926, when the respected South African statesman and polymath, Jan Smuts, wrote *Holism and Evolution*. Smuts, was, among other things, a philosopher and scientist who moved beyond the Enlightenment to introduce the idea of the 'holistic' development of organic systems. He was a strong critic of the 'struggle for existence' view of the world. He wrote that the supposed struggle is:

> an exceptional and not the usual procedure of organic Nature. This world is at bottom a friendly universe in which organised tolerant co-existence is the rule and destructive warfare the exception, resorted to only when the balance of Nature is seriously disturbed. Normally Natural Selection takes the form of comradeship, of social cooperation and mutual help. Normally also the organic struggle is very much in abeyance, silent effortless constant pressure of the physical and organic environment exercises a very powerful influence.

We have several reasons for using wholistic with the 'w' here. First, it links the human story to the wholistic perspective that existed before René Descartes launched the scientific era with his division between mind and matter. The chants of the druids, the epic tales of the Iliad and the Odyssey, the dream of Greek democracy and the creative works of the Renaissance all belong to a pre-science era where the whole was more important than the part.

Another reason for moving back to 'wholism' is the controversial nature of 'holism'. Among the various uses of the word, to be holistic was thought of as a claim to know all about everything. It could then, with some justice, be summarily dismissed as an impossible task. Alternatively, while opponents considered that it might have some merit, it was considered to be a lightweight idea with no solid foundation, an overambitious effort to assemble the sum of the parts into an indivisible whole.

When thinking about evolution, however, to be wholistic is none of these. Rather, it is to understand that it is the connections and cooperation, not the competition, between members of biotic communities that make up natural systems. Humans are therefore also to be considered in terms of their connections and interactions between their biological, social and individual places in natural systems. Finally, the adding of 'w' is a reminder of the central theme of this book, that connectivity and cooperation is the essential core of evolutionary change.

Scientific and public interest has moved on from the limited truths available from a narrow focus to seeking wholistic descriptions of the biosphere. James Lovelock's concept of Gaia as a single self-maintaining entity has been marked by its enormous popular appeal. Understanding Gaia in terms of its interactions and, at a lower level, understanding human constructs by exploring their relationship with Gaia now have priority. Scientific explanations using narrative and metaphor are of greater importance than the pursuit of detail.

Earlier, we tried to show the choice was not between reductionism or wholism, but that both had valuable contributions to make to ideas of how the world worked. Our metaphor for wholism proposes that understanding the world means taking the once-separated strands of knowledge and weaving them into the network of connectivity with which Gaia invests the planet.

Evolution: The Modern Synthesis appeared in 1942. This great work by Julian Huxley was a indeed a masterpiece of synthesis. It was, at the same time, a hymn to the success of the reductionists *and* a paean in praise of wholism; that *and* this – note the italics. It justified both approaches to the world by bringing together reductionist evidence from a wide range of different biological disciplines, including genetics, embryology, ecology, biogeography and taxonomy, and showing how each, taken singly, sheds light on aspects of evolution, and taken together reveal it as the overarching concept of life on Earth. It is probable that as many as half of all professional biologists were sceptical about Darwinism before the *Modern Synthesis*. Five years after publication it would be fair to say that fewer than 1 per cent of biologists had doubts about its validity. The *Modern Synthesis* had become the handbook of neo-Darwinism.

Even the basic role allocated to DNA changes when it is considered in its wider context. In fact, it is of such wide appeal that it has now become a popular catchcry, an acronym that appears to explain much but, in popular use, wildly distorts the original meaning. Thus, assertions like 'it isn't in his DNA to do that' or 'the ability to play the piano is in my family's DNA' are gross distortions of DNA's real role. DNA is like a carrier pigeon, but that is all. No-one expects the bird to act on the message it is carrying. DNA is only the carrier of the evolutionary message. Other molecules do the work.

To understand inheritance, we need to understand the method by which every cell gets its ration of DNA and the context in which it works. We have already noted that each human genome contains about 20,000 genes. Noble (2006) used the brilliant metaphor of the musical organ to describe DNA's role in inheritance. This was an excellent choice since the world's largest organs have close to 30,000 pipes, not too many more than the number of genes in the human genome. Together, the pipes can be made to play every piece of music ever written. Together, genes can be made to create every human being as a unique entity.

The pipes cannot, however, initiate the music by themselves, just as DNA does not initiate the living organism. An organist is needed to convert the musical score to a unique piece of audible music. The organist is expected to play the notes in the required context, and with specified order, timing and emphasis, although circumstances can change their interpretation. The juxtaposition and context of the notes can lead to a cheery tune or, less often, a masterpiece.

Within the cell, an array of different molecules interact with the DNA code and follow the instructions on how it should be 'played' or, as it is more usually put, 'expressed' in a living organism. The molecules within a human ovum, say, that interact with DNA, will build the human being according to specified order, timing and emphasis.

Who was the composer? The composer who wrote the musical score for the organ to play need not be the organist, but someone perhaps far away both in distance and in time. The composer of the DNA 'score' was natural selection itself, and it started its composition billions of years ago. To work the metaphor to death, natural selection has created billions of 'tunes'; these are the different species. Some have been briefly successful and then died out, while others, the so-called living fossils, have persisted without much change for up to half a billion years, and still others continue to be reinterpreted at every generation.

A very important characteristic of the genetic code is that it provides context. DNA is wrapped up into packages called genes. The genes are arranged on chromosomes. They do not, however, necessarily stay in the same place with respect to other genes. Cell division moves them about, and there other mechanisms, such as 'jumping' genes that move from one chromosome to another. The way these genes are expressed depends on the other genes alongside which they find themselves, that is, their context. Context is another way of saying 'connected with'. The connection may not be an actual physical one, but one mediated through other molecules. It is a wholistic system; without understanding the connections one cannot understand inheritance.

The recognition of wholistic systems continues beyond the genetic code. With the advent of electron microscopy, Altmann's bacteria-like granules turned out to be energy-producing ancient cells modified and living *within* a parent cell. Altmann's observation was an inspired anticipation of the Margulis symbiotic theory, put forward 80 years later, of the origin of modern cells as a collaborative enterprise between ancient cells. Like all good ideas, there had been early suspicions that such symbiosis might be the case; in particular, insights gained from studies of lichens, whose fungal cells are accompanied by symbiotic green cells that once were primitive blue-green algae. Margulis's work in the 1970s was a return to the study of the whole, and for 20 years was greeted with scepticism. Her insight, however, led to the rediscovery of the cooperative events that underpin multicellular life throughout the living systems of the planet.

Our image of the biosphere today is of the great, three-dimensional, network of connections that we can now call Gaia, a widely accepted, appropriate title for the interrelationships between humans, animals, green plants, ancient cells, oceans, atmosphere and geosphere. Together these add up to a self-organising entity. This is not so different from the views of the ancient Greeks, who believed that they lived in a united world, at the mercy of the gods of the sky, the land and the oceans that they personified as Zeus, Demeter and Poseidon. The idea of connectedness is also inherent in the aphorism beloved of the old alchemists 'As above, so below'. It originated in the Vedas, the ancient Hindu scriptures, as 'that which is Below corresponds to that which is Above, and that which is Above corresponds to that which is Below, to accomplish the miracle of the One Thing'. It implies that whatever happens on any one level of reality also happens on every other level. William Blake's *Auguries of Innocence* tells the same story:

> To see the World in a Grain of Sand,
> And a Heaven in a Wild Flower
> Hold Infinity in the palm of your hand,
> And Eternity in an hour.

Aristotle wrote that the totality is not, as it were, a mere heap, but the whole is something beside the parts, and was echoed by Euclid's Axiom number 5, 'the whole is greater than the part'. Leonardo da Vinci urged all to learn how to see – to realise that everything is connected to everything else – and Shakespeare noted 'one touch of nature makes the whole world kin'. In 1854, Chief Seattle, of the Suquamish People, was reported to have given a dignified address on the subject to the Europeans who were settling in his territory in the Puget Sound. The quotation that follows was actually 'recreated' much later for another purpose:

> The Earth does not belong to man; man belongs to the Earth.
> All things are connected like the blood that unites us all. Man did
> not weave the web of life, he is merely a strand in it. Whatever he
> does to the web, he does to himself.

To make this concept absolutely clear, Aristotle, when proclaiming the Greek equivalent of 'the whole is more important than the sum of the parts', was actually referring to Greek theatre. The whole play has an integrity and meaning, one that makes an impact on the audience that is independent of the contribution of any particular actor. Agatha Christie's famous play *The Mousetrap* has been running for 60 years with innumerable changes of cast and venue, yet its impact on a new theatregoer is not diminished. This wholist metaphor holds true for any cooperative venture. When the relationship is truly cooperative, then members make their own contributions to a meaning that transcends their own participation, to create a new and different and more potent whole. A great enterprise may continue for 100 years in a steady state, continuing its business even though the staff turns over many times.

Elinor Ostrom (2000) received the 2009 Nobel Prize for establishing the economic and ecological advantages of working with the whole of a community. She demonstrated that when natural resources are pooled, the rules for managing those resources evolve to make the system sustainable – provided that decision-making is transparent and democratic. The United Nations and the World Bank share these wholistic ideals, however difficult their realisation.

Reduction to parts still provides a useful tool in a wholistic enterprise, since science progresses by accretion of knowledge. Understanding of the whole becomes more complete as more of the unknowns become known. On the other hand, in contrast to reductionism, wholism is the idea that every whole has emergent properties, and that these cannot be predicted from knowledge of the parts. In another context, Donald Rumsfeld, US Secretary for Defense, put it less than succinctly in 2002:

> There are things we know that we know. There are known unknowns. That is to say there are things that we now know we don't know. But there are also unknown unknowns. There are things we do not know we don't know … And each year, we discover a few more of those unknown unknowns.

Even a quick look at the evolutionary history in Chapter 3 suggests that the evolutionary process is built on a series of emergent phenomena. Although emergent phenomena are commonplace, we rarely see them as such. Mistakenly, we tend to look at a whole as a static system – a reductionist view – whereas in a dynamic system, wholes are consequences of the influences of other wholes. Recurrence of a different spring in the northern and southern hemispheres is a consequence of the spin of a planet with a tilted axis in a solar system. Individual plants and animals are consequences of a fertilised ovum. Humans are the consequence of a particular pattern of DNA expressed in a social and physical context. Microscopic examination of either an unknown seed or a fertilised egg gives no clues to its final destiny. WB Yeats wrote:

> O chestnut tree, great rooted blossomer,
> Are you the leaf, the blossom or the bole?
> O body swayed to music, O brightening glance,
> How can we know the dancer from the dance?

The nut that is planted becomes the 'tree' and the 'tree' is all of the things listed by Yeats in the one living form. One cannot know the 'dancer', nor the 'dance', by simply studying the choreography. 'Dancers' may be changed and the new ones master the same choreography with different interpretations, while the dance itself will be influenced by different settings.

Darwin was a meticulous field observer with a formidable mind, but even he was not the first to put forward a theory of evolution as an explanation of the whole of life. His grandfather had done so before him. And, quite independently, so did Darwin's contemporary, Alfred Russel Wallace. Their views of evolution derived from close observations of the natural world, as well as from personal reflection. Darwin, from first principles, deduced that natural selection depended at least as much on collaboration among living things as on competition. Darwin clearly saw the golden thread, the connectedness of things. We have remarked on this before and make no apology for repeating it; at the end of the *Origin of Species* he was able to describe this marvellous vision in his 'tangled bank' paragraph.

In the second half of the twentieth century, the wholistic approach, the idea of cooperation among contributors that created something unpredictable, was finally coming into its own. The newly named field of ecology provided a broad, multidisciplinary and theoretical base for studying environmental systems that eventually produced James Lovelock's concept of Gaia as a system of interacting cycles controlled by feedback and feed-forward activation and inhibition.

Just as the reductionist interpretations of new information were influenced by the social context of their times, as discussed in Chapter 4, so too with wholism. The treatment of physical and social phenomena as sets of unrelated data changed to a search for ways to explain the same information within an interconnected system. Darwin's proposal of natural selection as the basis for the origin of species is a perfect example in itself. He gave us the prototype, the 'tangled bank', for a wholistic explanation that influences us still.

Even in Darwin's own time there were alternatives to the Establishment's hierarchical thinking. The group known as the romantic poets provided a wholistic counterpoint to the prevailing technological perception of the world. John Keats's 'trees young and old, that do a cooling covert make' gives the reader a very different image from trees cut down to fuel the Industrial Revolution. William Blake's 'dark satanic mills' and 'a robin redbreast in a cage puts all Heaven in a rage' symbolise their oppressed workers and, later, Charles Dickens's starving Oliver, in *Oliver Twist,* who asked for more, awakened the public consciousness of the terrible social impacts of the Industrial Revolution.

Even Darwin was concerned by the absence of fossils providing links between major groups of animals, although he confidently predicted that they would be found. Having been told to look for them, palaeontologists have been finding missing links in every evolutionary line ever since. The missing links in human evolution proved to be a consequence of too few hunters of human fossils on the ground.

Statistical theory advocates a certain care in interpreting data. The first Neanderthal human skeleton was discovered in the 1850s. If you have a sample of one out of a large population, then it is most likely representative of the average, because by definition the average is the most common form of that population. Sadly, a single observation was wildly out in this case. The Neanderthal individual was hunched, distorted and ogre-like. The skeleton later proved to be from an old, bent and arthritic man and far from the average for the healthy hominin. Unfortunately, in the public mind it became the model for all Neanderthals, and in the hands of creationists it became the distorted image of a kind of monster that, like 'nature, red in tooth and claw', reverberated down the years. Since then, many 'healthy' Neanderthal skeletons have been discovered. The richness of the human fossil record compiled over the last 50 years has finally dispelled the monstrous image left over from the nineteenth century.

The twentieth century move from a reductionist to a wholistic perspective changed our thinking in many significant ways. Behavioural studies had reached their reductionist zenith by mid-twentieth century. After that, there was a welcome move away from BF Skinner's view that animal behaviour could be explained by 'operant conditioning'; the idea that if a random behaviour has a good outcome then repetition establishes that as a normal pattern of behaviour. Conversely, if such a behaviour leads consistently to pain then avoidance becomes the established response. By training through reward and punishment, dogs could be induced to attack their owners, and humans to reject ingrained behaviour. Operative conditioning was reductionist in the extreme. It was less than helpful in describing the role of dogs in human society; and even less helpful in assessing the potential learning capacity of human beings.

Of the three co-winners of a Nobel Prize for animal behaviour in 1973, Niko Tinbergen and Karl von Frisch seemed happiest with this sort of reductionist analysis. This is not really surprising as they first worked on the behaviour of wasps and bees. The third winner was Konrad Lorenz, who had a foot in both camps. He famously found that new-hatched greylag geese 'imprinted' on him in the place of their mother goose. This is now known as a common phenomenon among animals; one of us once drove a big white panel van alongside a lambing paddock and it attracted the fond attention of a couple of newborn lambs that mistook it for a very large mother, much to the farmer's annoyance as he had the job of reuniting them with their mothers. Lorenz went on to observe the survival value of this imprinting during the period before they learned to forage independently. In much the same way, instinctive processes of adult mammals, such as fight or flight, are superseded by the later development of their capacity to modify the response according to the environmental context. Humans have infinite possibilities in satisfying their needs but even bees have a little wiggle room in their food-seeking behaviour.

In the 1980s, the wholistic approach was at last coming into its own. Ecology, now an established discipline, had started looking at human ecology and social systems in the evolution of societies. The legitimacy of including social, ethical and artistic evidence as well as the biophysical was clearly demonstrated by Lewis Mumford's 1953 *The History of the City*.

Wholistic ways of reinterpreting the outcomes of technological change include studies of hard (biophysical) and soft (social) systems. Reductionists are still contrasting the two in the twenty-first century, decades after the originator, Peter Checkland (1999), deplored their inability to put his then-new ideas of soft systems methodology into practice. Urban and environmental managers have largely moved away from regulatory responses to environmental change. Urban planners now include the needs of neighbourhoods and 'common ground' as well as 'roads, rates, and rubbish'. The current practice of adaptive environmental management brings together biodiversity, ecological systems and human livelihoods.

As an antidote to the extreme specialisation of the past, multidisciplinary teams were formed in order to deal with complex problems. When this simply continued the reductionist practice of separate contributions from the usual disciplines, the next response to challenging problems was to become transdisciplinary. Other ways of knowing, in addition to the formal disciplines, were included in policies for administration, research and education. Community, expert and organisational knowledge gave fresh insights into different aspects of the same events.

The use of the integrative word 'sustainability' came to the fore with the realisation that humans were responsible for environmental disruptions that put their own future at risk. The opening of the protective ozone holes at the planet's poles let in the Sun's UV light and exposed the entire human population to the possibility of UV-induced cancer. This time there was effective response: international cooperation led to the banning of the causative agent, the fluorocarbons. The lesson has been learned. Look about you: children's playgrounds are covered in shade cloth; chemical sunscreens claiming sunscreen factors of 70 are common (but check the validity of the claim). Even more important was the lesson about the close relationship between our planet and the rest of the universe.

Unfortunately, the same cannot be said of the reaction to the overproduction of carbon dioxide from burning fossil fuels. Here the public outcry at the risk of global warming from the atmospheric increase of the gas has been politicised. The dangers have not yet been enough to counter the influence of the industrial fossil fuel lobby, although the bushfires in Australia and elsewhere in 2019–20 have dented the granite facade of political climate change deniers. The people understand, even if their so-called leaders do not. The response of the latter has been to go on as before: more of the same (more coal and oil), delaying the enaction of a collaborative international response for abandoning coal, reducing gas emissions and investing in renewable sources of energy. It is, however, comforting to see that the private sector is responding well. They have read the evidence and self-interest is at work.

Humans now have responsibility for many of their own evolutionary selection pressures, a circular process that was not apparent in Darwin's own time. It has been recognised that a wholistic response requires social as well as environmental change. Yet another step towards a wholist approach to complex problems was the work of two social planners, Rittel and Webber, who labelled problems that had no current solutions, and required changes in the society that created them, as 'wicked'. Wicked problems were not necessarily moral problems; rather, they were wicked in the level of difficulty for their solution.

Rittel and Webber point out that solutions to wicked problems require responses from multiple interests, and so can have no single answer. Nor can they have a final answer, since social change brings fresh changes in their wake. The opposite of wicked problems are tame problems, problems that can be addressed through simplified single-factor responses. An example of a wicked problem is the need to balance the planetary atmosphere so that we can maintain the existing conditions for life on Earth. The simplest and most obvious step is to legislate to reduce the carbon dioxide emissions from industrial processes. The real dilemma is how to address all the aspects of a society that led to overproduction of the gas in the first place.

Human evolution continues, even though the selection pressures we once enjoyed have been radically altered by the changes we have made to our own circumstances. We have become unknowingly responsible for our own evolutionary direction and we are now in previously untrodden territory. Whatever our decisions, the Darwinian process of natural selection will continue just the same. The difference is that humans can circumvent environmental change, and even predict – to a certain extent – where and what will happen. Typhoid fever and diphtheria, outcomes of the crowding in the new mega cities, were defeated with the introduction of public health measures, clean drinking water and the use of carbolic soap. Mosquito-borne malaria was greatly reduced through use of insect repellents, changes in dress and the use of mosquito nets at night. Prevention, however, still escapes the experts; remember, the mosquito and the malarial parasite evolve too!

We are left on the horns of a dilemma. We can continue with our familiar ways of living on the Earth, ways that have themselves induced the significant problems we face, or we can take control. We can take control by identifying and acting on new ideas appropriate to life on a finite planet. This means taking advantage of the initiatives emerging in the move to a wholistic perspective.

Consider one of the foundations of our lives: how we choose to produce our food. A flood of information is emerging on the health consequences of long-standing land management practices. High usage of herbicides, insecticides and chemical fertilisers has been found to threaten the nutritional value of foodstuffs derived from monocultures of crops and cattle. Treatment of local grasses as weeds, stall-feeding cattle and the use of growth stimulants further impair the extent to which natural systems can restore the balance. Under conservative social pressures, farmers cannot easily abandon these now-traditional ways of managing the land. Drought degraded land and farming to the point of bankruptcy, and is considered responsible for an increasing rate of farmer suicide.

There is an alternative. Charles Massey's 2017 book, *Call of the Reed Warbler*, documents a revolutionary change from what he calls mechanical agriculture to regenerative farming. Farming practice is changing to allow natural systems to retake control, this time with encouragement from the farmer. This form of farming takes into account the whole of the landscape and the whole experience of living in that landscape. Personal accounts of the experience of taking up this challenge report increases in incomes, markets and nutritional levels of produce, and improved work-life balance.

Massey's conclusion is that although the case for regenerative farming has already been made, a different way of thinking is needed to put it into widespread practice. What is required are minds open to new practices, social and environmental, sensitive to relationships, and expecting and embracing the inevitability of evolutionary change. He is not alone. Meadows et al., in *Limits to Growth* (1974, updated in 2013), are helping to change the world's thinking about the future. Meadows's recommendations are to start with the vision, be open to any path by which the vision can be realised, be patient and persistent and be true to the vision. Humanity is at a great turning point in its history, perhaps as great as the discovery of fire or the invention of the wheel.

A continuing issue for which no solution has yet been found is the heating of the Earth's atmosphere and oceans. Separate government agencies, education departments, research teams and community advocacy groups have developed different approaches: carbon dioxide reduction, pollution control, policy development, change management, maintenance of biodiversity and the rest. Work by one of the present authors, Brown, (Brown and Harris 2014) on wholistic decision-making has explored the wicked problem that still exists, even in a fading reductionist era, of establishing connections between the social structures and courses of action. Single disciplines still maintain monopolies on their own forms of knowledge about education, and their applications in the professions.

An early effort to combine the decision-makers involved bringing together as wide a group as possible in workshops. At each workshop, the participants were in search of whole-of-community change. In time, there were more than 300 of these workshops spread across five continents. A pattern began to emerge. Contributors to the workshops brought, separately, to the discussions individual, community, specialised, organisational and wholist ways of approaching a shared issue. Each of these ways had its own objective, language, timetable and knowledge store. They were so distinctive that it seemed reasonable to describe them as each being a knowledge culture of their own. When placed together in the same workshop, members of a community could be persuaded to talk to one another, and even collaborate on responses to a shared issue. In this way, wholistic programs on one issue were expanded to include a wholistic understanding of environmental and social change.

Midway through the workshops it became apparent that there was another distinct subculture of members. These were often the instigators and/or the drivers of the wholistic programs. They might have come from any one of the knowledge cultures, or they might have been independently minded. In each case, they drew on different, though overlapping, sets of knowledge from their own knowledge cultures. These individuals were reflecting on the biophysical, social, ethical, artistic and sympathetic aspects of the issue for themselves.

The notable success in arriving at a team response to a wicked problem was obtained by treating all workshop members as individuals, rather than as representatives of their own specific knowledge base. Continual feedback through mutual dialogue led to mutual learning among the members. It brought together the full range of available evidence on each of the issues, so that each member could contribute equally.

The next question, then, must be, where does all this activity sit with respect to natural selection within and between human populations? We know that the pressures on the human species are unprecedented, and that the problems it faces call for wholistic solutions. The question is, of course, much more difficult. While we can describe what has happened, and what is happening to the species, the present is so complex an interaction between multiple factors that what will happen in terms of natural selection requires a crystal ball! Nevertheless, in Chapter 13 we attempt a lighthearted look at a future, and identify some of those pressures and the possible responses.

7

ENERGY: WHERE
IT ALL BEGINS

In which the relationships between energy and evolutionary change are followed from the first ancient cell to Gaia.

The continuous flow of energy from the Sun is responsible for the maintenance and self-organisation of life on Earth. Cycles of energy establish the environmental conditions under which evolution can occur. We can follow the interrelationships between energy flow and evolutionary change from the very first cells to the complexity of the natural systems of today.

'The flow of energy through a system tends to organise it' is an aphorism that describes the world that we encounter every day. Provide human babies with adequate nutrition and, with the energy they derive from it, and within the possibilities of their genetic program, and with culturally appropriate care, they self-organise into adults. Here we address the processes of self-organisation of Gaia, and its many characteristics, that are brought about by the flow of energy from the Sun. These include the amount of time that energy spends circulating through the biosphere (its dwell time), how it engenders cycles of resources and how it supports increasing complexity. All this is achieved in the face of the tendency of things to run down, the tendency for disorder. Without energy flow, in the words of WB Yeats:

> Things fall apart; the centre cannot hold;
> Mere anarchy is loosed upon the world,

Energy flow from the Sun is an absolute requirement for all life on this planet (with the exception of a number of ancient cell types that can use geothermal or chemical energy). The Earth absorbs the energy like a huge greenhouse, with an atmospheric shield of gases instead of glass. Inside the 'greenhouse' every animal depends ultimately on the ability of green plants and algae to turn carbon dioxide into sugars and release oxygen. This is the chemical process of photosynthesis that takes place in the small specialised organelles in the plant cell called chloroplasts whose ancestors were once, long ago, independent ancient cells. It is estimated that the humblest of ancient cells, the blue-green algae that comprise a major part of marine plankton, are still responsible for 20 per cent of the total photosynthetic output of all green plants. This by-product of photosynthesis, oxygen, leaks into the environment and maintains our stable, breathable atmosphere.

Figure 15 is a diagram of the Solar System as a steady state. The Sun radiates energy in all directions; some of that energy is 'captured' by the Earth. It flows through the biosphere and is finally radiated into space. As long as the entropy – let's think of entropy as disorder or chaos – as long as the entropy of the whole system increases, it does not violate Newton's laws if a small corner of it – the biosphere – shows a decrease in entropy and becomes more ordered or complex.

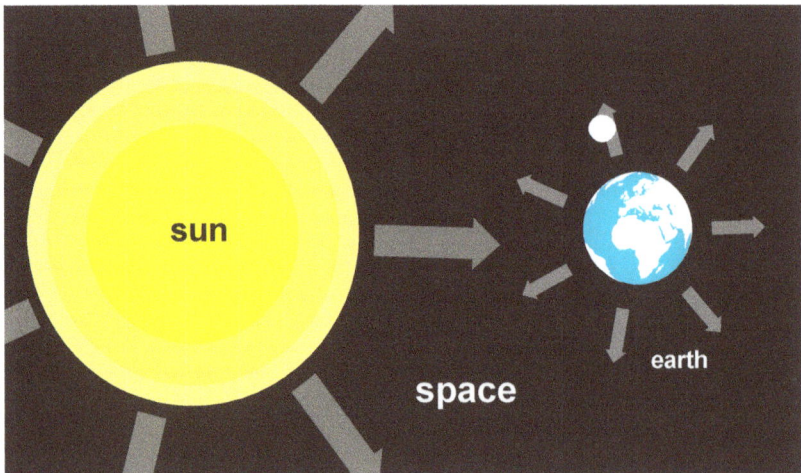

Figure 15. The flow of energy in the Solar System.

The relationship between the Sun and the Earth is therefore one of energy flow and, from the short-lived point of view of living things on Earth, to all intents and purposes, it is a steady state, a condition that does not vary with time. Certainly the Sun's output has fluctuated, and will fluctuate, but the time scale for such fluctuations is measured over millions of years: from our point of view, that's steady.

Energy therefore flows from its source, the Sun, whose light shines on the photosynthetic green mantle of Earth. The sugars that are created by green plants are used immediately to sustain their own lives and growth. Any surplus is laid down in storage organs, such as roots and tubers, for future use. It may, however, be stolen by other organisms, including humans buying potatoes and onions at the greengrocer.

Plants and animals die and their remains become food to be converted into energy by other plants, animals, bacteria and fungi. After a little while, they also die and become humus, the dead organic matter on which future plant growth depends. Sometimes, when environmental conditions permit, once-living organic matter is fossilised as coal and oil deposits. Finally, excess energy, as heat, is radiated into the unfillable 'sink' called space. The amount of time that energy spends moving around Gaia is called its 'dwell time'. It is the time that the Sun's energy may be available for use by living things.

In spite of the efforts portrayed in numerous science fiction stories, it is not possible to imagine where the sum of evolutionary change will take us under these circumstances, even 100 years hence. For one reason, ever since the discovery of fire, an important component of the total human energy budget has been provided by fossil fuels laid down millions of years ago. We do not know, though we can make informed guesses, the form of energy that we will need in the future. The present choices, as we turn away from fossil fuels and toward renewables, seem to be nuclear fission, solar, wind, hydroelectric, geothermal, tidal or other uses of the omnipresent gravity. We do know that energy will be needed, however we decide to harness its flow.

We can best understand what can be accomplished with energy flow systems by considering the many applications of the concept of flow in human societies. For example, one of the most important innovations of Roman civilisation was their engineering of water flow and the treatment of sewage. The Romans probably take the credit for creating the first system of drains. They conceived of them as simple removal devices, to transport unwanted waste products somewhere else. They were not concerned with sewage treatment per se – they wanted to get rid of the smell, the 'miasma' that they thought was responsible for disease.

A modern sewage system involves treating the sewage and recycling the safe organic material that is produced. In efficient systems, from early agriculture to major cities, sewage material has been allowed to act as recycled energy. Medieval farmers knew this well, as their livelihood depended on their dungheap. In a well-organised sewage system, the dwell time – the storage time between the origin of the waste and the use of the effluent – is very important. Flow must be continuous and constant. If the sewage is allowed to build up, it will overflow in inconvenient places. If there is no flow through the system, it will stagnate and provide an energy source only for bacteria. Ideally, in a modern sewage farm, the rate of flow allows useful biological systems to operate; the heat they generate during dwell time eliminates the disease-carrying organisms and allows the formation of the energy-carrying gas, methane, which can be stored for immediate or later use.

Consider another flow system, common in medieval times, the water-driven flourmill. It was typically set close by a stream, at a point on the watercourse where the water coming from higher up had sufficient mass and velocity to do work as it fell downhill. By inserting a millwheel into the stream, the miller was able tap its energy to turn the wheel and hence turn the grindstone to mill the flour. The volume of flow was very important. If too little water passed down the stream it would not have had sufficient energy to turn the wheel. To guard against this eventuality in times of low rainfall, watermills have millponds above the wheel to collect and store as much water as possible during wet periods. By operating a sluice the miller can direct water from the pond into his millrace that channels the water past his wheel. For the miller, ideally, the dwell time for storing water in his pool should be as long as possible. He wants to be able to rely on a constant, effective flow to mill his grain.

These two examples demonstrate that in dynamic flow systems, like sewage disposal and milling, both rate of flow and dwell time are important. These ideas create a metaphor repeatedly encountered in human affairs. We talk of the flow of time, of music, of poetry – and of the wealth that makes all the aforementioned possible. Economists have a maxim that 'the flow of money through a system (a company, a village, a town, a city, a country) tends to organise it'. That is, it increases its structural complexity (see Fruth 2015). Rate of flow of money is obviously important; too little and an organisation shrinks as jobs are lost.

As employees, people represent money. The decrease in the rural population in England during the Industrial Revolution was due to farm hands migrating to the cities, where the 'dark satanic mills' of Blake were to be found. The flow of people was equivalent to an outflow of money from the farming sector to the wool and cotton industries. The remaining workforce was no longer sufficient to support the existing agrarian infrastructure. Even worse, in the society built around the Industrial Revolution, the dwell time of money was inadequate to allow much development for the working community. It flowed straight into the pockets of the great mill, mine and landowners who had other uses for it. More recently, as more complex flows developed, people have developed structures – cooperative societies, schools, hospitals – to increase the dwell time of money and address their different needs.

The flow of solar energy from the Sun, through the Earth and back into space is the embodiment of Newton's laws of thermodynamics reworked as Carnot's principle (see, for example, Laranjeiras and Portela 2016). Heat always flows from a hotter body to a cooler body. It was implanted indelibly into young minds nearly 60 years ago by one of the songs of Flanders and Swann that began, 'Oh, you can't pass heat from the cooler to the hotter'. (To underline the power of such a mnemonic in teaching children, one of us has never forgotten that the operating cycle of the internal combustion engine is 'suck, squash, bang, blow'!)

During the process of energy flow from hotter to cooler, Newton's second law demands that entropy, the amount of disorder in the whole system, must increase – remember, 'things fall apart'. Entropy is a technical word but perhaps it is easier to think of it as the universal and continuing breakdown of structure. Energy collected from the Earth's self-organising systems is continually flowing outward, as heat, into space. It follows that a highly organised system, like a human being, with lots of order created out of information contained in DNA, is relatively low in entropy (disorder). In creationist literature, it is often remarked that, because order increases and entropy decreases in living organisms, rather than the other way round, the second law does not apply to them. This, they claim, somehow sets living things outside of normal thermodynamic considerations. This thinking demonstrates a fundamental misunderstanding of energy flow principles. In steady state systems as long as the entropy (disorder) in the whole system increases, then, it is quite allowable for small parts of it to show less disorder. This is the reason for the entropy decrease achieved by Gaia within the background of the increasing entropy of the Solar System. The small planet uses the dwell time of the energy for the self-organisation of living systems. Figure 16 illustrates the great carbon cycle that drives Gaia.

If the Earth were a homogeneous, smooth ball of rock, it would warm up uniformly and cool uniformly. The real Earth is very different, it is composed of trillions of objects, some of which warm up and cool down more quickly than others. In addition, the Earth rotates, presenting a continually different surface to the Sun. Instead of uniform warming it is patchy, as the hottest objects pass heat to the more cool ones. All these interactions combine to increase the dwell time of energy on the surface of the Earth, giving it time to do work. Green cells and plants are among these trillions of objects and they capture energy and process it in very small, manageable parcels indeed.

Perhaps an analogy will make this clearer. The Sun pours out energy unremittingly. When it hits the Earth the only thing that happens in the absence of life is that everything gets extremely hot and then cools down as the Earth turns, rather like conditions on the Moon. This is what is meant by a vertical temperature gradient.

Let us now suppose that the flow of energy is like the flow of a great river. Imagine this river going over a waterfall with a 100-metre drop. The potential energy of the water as it flows over the edge is very great but much of it is dissipated at once when it hits the ground, gouging out a great pool at the bottom in a cloud of spray, and then flowing on its way. The fall of water is also a vertical gradient.

Now, instead of the water dropping 100 metres, imagine that it is flowing down a staircase of 100 steps, each with a height and width of one metre. The vertical gradient has been replaced by one of 45 degrees. The energy lost by a 1-metre fall is not very great at each step, but 100 steps, one after the other, adds up to a great deal of energy indeed.

It is now time to reintroduce the water-driven mill and a miller who mills flour. He's a resourceful chap who knows about gradients. He wants to harness the energy of falling water to turn his grindstone. He immediately rejects the idea of trying to tame a waterfall because the vertical gradient

– the long drop – produces more energy than he can reasonably handle; his millwheel would be reduced to matchwood. He looks for a shallower gradient along the river, and very soon discovers a promising site. But it's still quite a large river just there and he imagines what it must be like in the wet season, in full spate, probably bringing small boulders down that will damage the millwheel he is planning to build. He therefore decides to position his wheel some way from the river. So, from a point upstream, he constructs a small canal and a sluice gate that collects water and discharges it into the large pond – the mill pond – that he has dug ready. By opening a sluice gate in the pond he can release water into a channel that runs right by his mill where he intends to insert a millwheel. The water will turn the wheel that turns a grindstone that will grind his corn and he's in business. The water has less energy as it leaves the millwheel, but that is still a lot of energy going to waste if someone wants to tap it. Being an ingenious fellow, he develops a clever system of gearing to trap some of the surplus energy from grinding flour and adds a potter's wheel and, later, a lathe. He can now make and sell kitchenware and furniture as well grinding flour for the local village.

And that is analogous to the thing that all living green plants do. They capture some of the energy from the Sun and use it drive a series of small chemical steps, a metabolic pathway, to make simple sugar molecules.

Life thus deals with the vertical drop of the thermodynamic gradient from the Sun to the Earth by converting it into a 'descending staircase' with many chemical steps. Each step represents one tiny, manageable bit of energy. A plant captures energy with its chloroplasts and uses it, by a series of reactions, a metabolic pathway, to make sugars and starch. These might be stored in specialised organs, such as potato tubers or daffodil bulbs. When energy is required for respiration and growth by the plants, another pathway is brought into play and starch is first turned into glucose, that is then oxidised (or 'burned' as a marathon runner 'burns' carbohydrates) to provide the necessary energy. During the process glucose gives up its energy in very small parcels, rather than all at once. This is just as well, because if it did give it all up at once it would go off like a firework and cook the cells in which it was happening. Starch dust so readily generates heat when it is breaking down that sometimes it blows up wheat silos.

An industrial power station does essentially the same thing as the local miller. The power station burns coal to carbon dioxide and water and energy is captured and converted to a different form, electricity, for use.

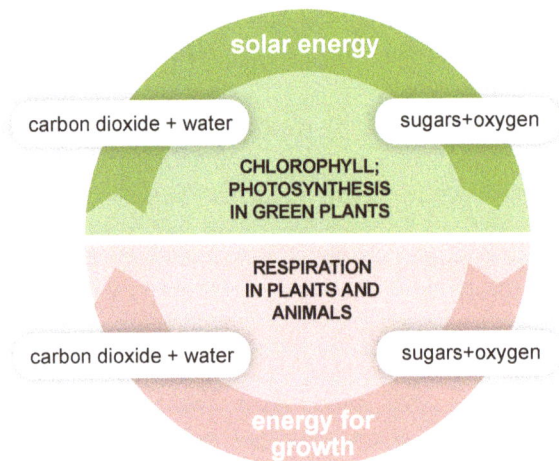

Figure 16. Gaia's carbon cycle.

The great energy cycle, the carbon cycle of Gaia, that supports all life on Earth is thus completed. What has happened is this:

1. Energy capture from the Sun by green plants and blue-green algae:
   - **carbon dioxide** and **water** are converted by photosynthesis to simple **sugars**, and **oxygen** is released. (This is where all the oxygen in the atmosphere comes from.)

2. Respiration by all living things – animals, green plants, fungi and ancient cells:
   - **sugars** are oxidised by **oxygen** and converted to **carbon dioxide** and **water**.

Very little observation will demonstrate that living things increase in complexity during evolution, from ancient microbes to modern blue whales. Increasing size and complexity are only possible because energy flows through the biosphere. Complex systems are maintained by energy flow, so it is obvious that achieving greater complexity uses up greater amounts of energy. The limit to complexity is defined by the maximum amount of solar radiation consistent with the continuation of life on Earth. We cannot know what point on this upward curve of complexity we have reached. It seems possible that the rate of increase in complexity is slowing. This does not mean that the overall rate of evolution is necessarily slowing, as we must also presume that it continues simultaneously at all levels of organisation from the single-celled microbe to multi-celled blue whales.

If energy flow generated by green plants leads to greater and greater complexity, it must also lead to increasing instability. More energy capture is required for maintaining increased complexity.

As life gets 'pumped up' to higher and higher levels of complexity and connectivity, it still remains hostage to chance. Asteroids colliding with Earth, or a period of increased volcanic activity can raise dust and vapour clouds that reduce the amount of the solar radiation that reaches the surface of the Earth. In the past, these have resulted in wholesale extinctions. Six major extinctions are on record since the earliest emergence of life. Another possible candidate for the status of great extinction is the current stripping of natural habitats by humans. On the upside, an extinction provides great opportunity for evolutionary change, since there are then a multitude of potential environmental niches left vacant.

Lovelock must welcome the modern emphasis on the role of niche improvement in evolution, since his Gaia hypothesis is predicated on just this concept. Niche improvement goes on all around us all the time. Any gardener will tell that you that planting potatoes improves the soil by breaking it up and improving drainage. The gardener will then take advantage of it by planting other vegetables in the bed the following year. It is a trivial example, perhaps, but now think about rainforests, which, over thousands of years of evolution, have fixed carbon dioxide into leaves that fall and decay, creating humus that was not there before the rainforest. Or birds that build elaborate nests that permit the more efficient incubation of eggs; or the nests of wasps, the hives of bees, the mounds of termites. Or the first green plants to invade the land with their probing roots that first assisted weathering of the rocks and later, much later, stabilised the valuable soils that came into existence or *Homo sapiens,* the greatest niche 'improver' of all animals. Think of the global cities, households, hospitals, schools. All of these social constructions require reliable flows of energy to establish, maintain and improve the *Homo sapiens* niche.

It is all part of a great, if unintentional, collaboration. The changed environments of each niche are hotbeds of natural selection. An improved niche is a territory in which the mix of energy flow, geochemistry, microclimate and coexisting biology becomes capable of supporting forms of life beyond that for which it was created. In the same geographic space that contains, say, an ant's nest, there are countless other niches that are occupied by non-ants: plants and animals that peacefully coexist with the ants. They would not exist had it not been for the ants improving their own fitness to survive by building the nest. All of them interact and are subtly changed by the interaction. Leonardo da Vinci wrote: 'realize that everything connects to everything else'. Gaia is a great all-inclusive network of niches. Together, niches form a huge and intricate three-dimensional mosaic that sits on the molten core of the Earth.

It seems possible that modern society is approaching a crisis point in its handling of energy, too clever by half in handling its own niche improvement. Unlike the ants, humans can visualise in advance ways in which they might work to improve their lot. No ant ever made a wooden armchair, although termites, no doubt, have eaten many!

As the human population grows, energy flow from the Sun will remain relatively stable. The world's ecosystems, however, are being exploited at an ever-increasing rate. The World Wildlife Federation has recently warned that, by the middle of this century, humans will need the equivalent of two planet Earths to continue to live as we are at present. Simply put, we are using the planet's resources more quickly than they renew themselves. If there is less to go around, standards of living must fall or else human populations become ever more sharply divided into the haves and have-nots. Australia is a test case. We have limited resources and what we do have a lot of, like iron ore and coal, other countries are wanting less and less. The human carrying capacity of Australia is limited by the availability of water and generally poor soils. Some authorities suggest that a human population in Australia of between 50 and 100 million is the limit without vastly improved techniques for agriculture and renewable energy use, and much greater investment in infrastructure. The real point is this: in a steady state solar system, energy flow is finite and not all the energy is available for use. One has always to pay one's dues to counter the tendency to disorder. Or, to put the relationship between Earth's energy balance and increasing entropy in the universe in terms that a gambler could understand (popularly, if inaccurately, known as Ginsberg's theorem):

> You can't win.
> You can't break even.
> You can't quit the game.

Whatever happens, you lose.

Gaia has three important characteristics that are essential for her continued existence, all of them involving constant energy flows. They are: first, great complexity and a high level of order; second, feedback and feed-forward cycles that regulate the whole system to give stability and longevity; and third, cycles of cycles – hypercycles. They are exemplified at the cellular level, in biochemical pathways that cycle; at the organismal level by the many interacting reproductive cycles; and at the environmental level by the carbon and nitrogen cycles. The planet itself cycles around the Sun.

Cycles underlie the whole of human existence. Humans are conceived, increase in complexity until they reach maturity, a steady state, lasting perhaps 30 or 40 years, and they contain many biochemical and physiological cycles. Humans, in fact, are rather like the blacksmith's hammer, the one that belonged to his great-great-great-grandfather. The shaft has been replaced half a dozen times, and it has had a few replacement heads, but it's still the same hammer – isn't it? All human cell types undergo renewal cycles at different rates, except the cells of the cerebral cortex, which are not renewed, although they can be reprogrammed. But you are still the same person – aren't you? After a time the steady state decays and senescence and death follow, when all human order is lost, and its remains are returned to the ground to participate in other cycles.

The evolutionary history of life on Earth is a history of the complexity and order of life, as they increase with geological time. We have agreed that as complexity increases so does the need for energy to maintain it. An ancient cell is less complex than an independent modern cell. These are less complex than multicellular organisms like sponges. Sponges are less complex than elephants. But it is important to remember that bacteria, protozoa, sponges and elephants are all with us today – and what is more, ultimately, we are all related, in that we share an ancient common ancestor. Each level of organisation does not therefore replace the previous one, rather it builds on what has gone before so that the whole of life becomes an interconnected, layered phenomenon.

To travel, in your mind, the path that leads from the simplest cell to the mind-boggling blue whale is to undertake a journey of continual wonder. Twice before we have referred to the shocking discovery of the apparent small size of the human genome and remarked how astonishing it is that it comprises only about 20,000-odd genes. Even a single-celled soil amoeba has almost 16,000, while a tomato plant has more than 30,000. Many of these genes are highly conserved; that is to say, they represent combinations of genes that remain responsible for the most fundamental processes of life. The fact that they are present in all animals, plants and fungi allows us to determine degrees of relatedness. From this we learn that we all speak the same genetic language and, as we have noted, DNA is a celestial Lego set, a vast collection of building blocks that are used over and over again in different combinations to create the many forms of life.

The Earth is about 4.5 billion years old. Some of the oldest fossils are found in rocks that are about 3.5 billion years old. They are thought to have been primitive cells (cyanobacteria, also called blue-green algae) that obviously must have had their beginnings well before that time. There are various scenarios that attempt to describe how the world before life brought forth the first 'protocells' that were their ancestors, but they all agree that a continuous flow of energy such as that from the Sun created the necessary thermodynamic preconditions. There are many scenarios that attempt to explain what happened next and a recent book by Ward and Kirschvink (2015) gives an excellent account of all of them.

Viruses are the first step in the journey to the blue whale. Recent discoveries indicate that the virus is, as it were, part of us. This is both scientifically and aesthetically pleasing, as it emphasises yet again that we are *all* linked into Gaia and we no longer might have to make exceptions for viruses. Perhaps it is not so surprising after all, as viral DNA is frequently found inserted into in the genomes of higher organisms and we can now legitimately embrace it as a source of evolutionary novelty.

On Earth, wherever there is an accessible energy source, a life form has taken advantage of it, and evolved from there. The discovery of volcanic vents in the deeps of the ocean provides the fuel for a number of 'origin of life' postulates. 'Black smokers' (Tunnicliffe 1991) attract attention because they create environments equipped with the necessary chemical set for early life and in which dwell representatives of the earliest living forms on Earth. Cooperative biology starts here!

It has even been suggested that, if recent evidence of ancient volcanic activity on Mars is correct, life may have had its origin there. Calculations that Mars rocks, thrown out by volcanoes, could reach escape velocity and fall to Earth offer the beguiling prospect that we are all Martians. Some billions of tonnes of Mars rock are supposed to have made this journey so far. Perhaps one was sheltering a primitive Martian molecule that won the jackpot by ending up near a vent.

However it happened, the first protocell (a cell that is not yet quite a cell) that appeared was essentially a membrane containing a large number of functioning chemical units (molecules). The formation of a boundary membrane was an absolutely critical step in organic evolution. Only by containing these molecules within a bounded environment in effectively high concentrations, together with an energy source, could there be any internal control. Homeostasis, that is to say, regulation of the composition of the protocell within limits defined by being alive, was at last possible. The final important step was the acquisition of an *almost perfect* mechanism for cell division. It must have been *almost perfect* because a perfect mechanism would always result in identical daughter cells. Subsequent evolution would then have to depend on external events, such as irradiation, to produce variation and to allow natural selection. Darwinian selection hangs on the fact that offspring vary, so the *almost perfect* system sets up the preconditions for evolution to take place.

Today, all the Protista (protozoa and single-celled algae), fungi, plants and animals are organisms made of modern cells. While we humans are also comprised of modern cells, it must be remembered that many other ancient cells are found within our intestines and upon our skins. These are our partners and most of them contribute to our wellbeing. Others, relatively few, are parasites that attack us and contribute to our illnesses and our mortality. We are actually doing pretty well because, according to one estimate, the world contains about 5 million trillion trillion (5 followed by 30 zeroes) ancient cells. This works out at a biomass of about 1,000 times the total weight of the human biomass in the world. Approximately!

Early attempts to describe modern cells as the result of symbiosis between earlier cells were laughed aside. It was no surprise, then, that the Margulis hypothesis that modern cells were made up of a collaboration between three or four ancient cells met with frank disbelief at first. The implications of the Margulis hypothesis, however, are enormous. Three billion years ago the Earth belonged to the ancient cells. Competition between them must have been intense. Cells had preyed on or parasitised other cells. Adaptations of offense were countered by ones of defence. While many cells must have succumbed to attack, many others managed to coexist with their attackers. But some billions of years ago a third type, working in intimate cooperation with other cells, had managed to establish many symbiotic associations, to be tested by natural selection. In support of this view, very recently, an intriguing new type of organism has been discovered which appears to be an intermediate form – a missing link! – between an ancient cell and a modern cell. This missing link, *Parakaryon myojinensis,* if confirmed, is compelling evidence for the capacity of cells to form a variety of cooperative relationships (Yamaguchi et al. 2012).

Every multicellular organism, including humans, is comprised of modern cells. On top of this, humans have a relationship with 500–1,000, at least, types of ancient cells, billions of bacteria living in the intestine and elsewhere. Bacterial cells are so much smaller than modern cells, and it has been estimated that, in the average human, there are three times as many microbial cells as modern ones. A human is truly a biological community.

And here is a thought. Among the earliest evidences of life on Earth, more than 3 billion years ago, are fossilised stromatolites. They are structures comprising microbial 'mats', especially blue-green algae, bound together in layers by concretions of sand grains. They grow as columns in warm shallow waters. A visit to Shark Bay in Western Australia is well worthwhile, for anyone interested, to see modern examples. The stromatolites are actually a scaffold on which these ancient cells could thrive in the light. The thought? Perhaps humans, from the point of view of their internal ancient companions, should be regarded as mobile stromatolites!

This chapter has moved from the stark realities imposed by the physical laws of thermodynamics to a consideration of the biophysical laws of energy flow and the 'steady state'. The apparently inevitable natural synthesis of the molecules on which life depends, followed by self-organisation to produce the first ancient cells, was followed by the most momentous event in evolutionary history, the collaborative symbiotic origin of the modern cell. Forms of life interacted cooperatively at the first possible opportunity. This created a precedent that continues to this day and has been going on for billions of years – perhaps even longer.

Humanity started its own journey to harness energy with the discovery of the uses of fire only 1 or perhaps 2 million years ago. The discovery that fire could be controlled led inexorably across the 2 million years by way of primitive technologies of the stone, bronze and iron ages to the Industrial Revolution. Along the way, the energy in wind and water has been also harvested, and tidal energies have been tapped.

The discovery of radioactivity opened doors that many think might better have remained closed. However, once a door is open it does not seem to be within the power of humanity to close it again – at least, not without first finding out what is on the other side. So far, the atomic bomb has been used in war on only two occasions and now, surprisingly, atomic power stations are in decline, not due to ethical considerations but to obsolescence and the lack of profitability in building new ones. This unexpected outcome is discussed in 'Nuclear Holiday' by Lisa Grossman (2017). Perhaps an era of utilisation of renewable energy sources beckons until fusion – using the same reaction that occurs in the Sun – becomes a viable alternative.

The chapter has pursued the many ways in which energy supply and evolutionary pressures go hand in hand. These include the length of time that energy can be stored (dwell time), the expansion of ecological niches, the cycles of resources that together make up Gaia, the universal tendency to disorder and the complexity of self-organising systems. Each of these are fertile sources of difference, and with difference comes the evolutionary opportunity.

The next chapter is concerned with the ubiquity of the evidence of cooperation between different forms of living matter. It starts from the point that parasitologists have long been aware of. There is no known single species of organism that does not have another one living on, in or with it. It moves on through evolutionary history to reach the evolution of the social-based system that is humanity.

8

EVERYTHING IS CONNECTED

In which we examine the evidence for cooperative evolution and show that everything is indeed connected to everything else.

In this chapter we argue, on the basis of evidence that we consider to be incontrovertible, that the predominating way of life on Earth involves the collaboration of organisms in a grand amplification of Darwin's vision of a tangled bank. We examine, with numerous examples, the diverse forms of cooperation – partnerships, symbioses, parasitism, multicellularity and human society – from ancient cells to human beings, that illustrate the weaving of Gaia's golden network.

The mutual dependence shown by living things is one of the defining characteristics of life. Symbiosis describes the cooperative enterprise in which all parties contribute to the wellbeing of the others in the partnership. It is a phenomenon that has been well known for centuries, but it is only recently that scientists have begun to understand its full import. Every school biology book is packed with examples of animals living in mutually advantageous harmony, representing them, quite justifiably, as among the wonders of biology. They are, however, only the more spectacular tips of a very large iceberg. It is, actually, a commonplace in biology.

Symbiosis could be defined as:

> the interaction between two or more genetically different organisms living in close physical association, to the benefit of all parties.

From a philosophical point of view, this is too limited, suggesting, as it does, that cooperation occurs only between different species. In fact, symbiosis also describes the arrangements between all animals that associate – insect colonies, herds of elephant, shoals of fish and flocks of birds. Hive and herd structures benefit the whole in dealing with predators. There is 'safety in numbers' in anthills and armies.

Closer to home the definition specifically excludes the relationship between a mother and her intra-uterine baby and, later, between a mother and her newborn. It excludes the whole of human society where, according to anthropologist Margaret Mead (1928) borrowing an African proverb in her famous but much-criticised book on adolescence in Samoa, 'it takes a village to raise a child'. Some experts in mammalian reproduction have gone so far as to describe the relationship between mother and foetus as parasitic. We reject this as it groups an essential relationship with inessential tapeworms and warble flies! The definition is one that clearly has problems.

A description of symbiosis as it occurs in the real world might be to delete the word 'genetically':

> the interaction between two or more different organisms living in
> close association, to the benefit of all parties.

Finally if we go back to basics, let us simply talk about cells, for the first modern cells are themselves tri- or tetra-symbionts:

> the interaction between two or more different cells living in close
> association, to the benefit of all parties.

We now have a definition that includes the whole of biological creation. Individual animals and plants become symbiotic associations of cells. Gaia herself is a symbiotic construct of all living forms, both plant and animal. It includes animals in mated couples and human societies.

Here are some well-known examples of symbiosis between genetically different organisms.

Mammalian ruminants (sheep, cows, deer and their relatives) could not survive without the microorganisms (ancient cells and fungi) that dwell in the section of their stomachs that is called the rumen. These microsymbionts have the ability to break down cellulose and, from it, generate essential nutritious metabolic products. One study of cattle showed that the rumen is the perfect incubator for *at least* 5,000 different 'operational taxonomic units' – species of bacteria, protozoa and fungi. This was determined using modern genetic techniques. Remarkably, the populations of the rumen inhabitants of individual animals were very similar, suggesting that each animal was regulating its own population in the same way (Jami and Mizrahi 2012).

Then there are the hindgut fermenters, animals with large posterior fermentation chambers, such as elephants and rhinos, horses and tapirs. There are the stomach fermenters, like the hippopotamus and some large marsupials. Thus, most of the modern mammalian megafauna (excluding the carnivores for the moment, although they also have interesting inhabitants in their guts), depend on intestinal fermentation one way or another. Almost certainly, so did the large herbivorous dinosaurs.

What worked once for termites long before mammals appeared on the scene can work again ... and again. Hindgut fermentation helps quite a few of the smaller mammals. Rabbits make up for having a small, rabbit-sized intestine by eating their own faeces. This process is called coprophagy. At night rabbits produce soft, green, partially digested faeces and eat them, giving the microbes in their intestine a second go at breaking down cellulose. Important nutrients are synthesised by symbionts in the posterior, large intestine while absorption occurs in the anterior small intestine. What else can a poor rabbit do but recycle? Coprophagy also happens in rodents and it has been observed in koalas, ringtail possums, piglets, foals, dogs and nonhuman primates. Pigs regard human faeces as an excellent source of nutrition!

Vertebrates, however, are a very small class in the biological scheme of things so let us widen the search. Algae have a truly remarkable record of living with the cells of other organisms and cooperating with their owners. Lichens (fungi with algal symbionts, either ancient blue-green algae as well as modern single-celled algae) are a whole class of organisms that would not exist were it not for an obligate symbiosis between a fungus and an alga (Nash 2008).

Algae live in the cells of the reef-building stony corals; the Great Barrier Reef would not have existed if it were not for them. Coral bleaching occurs when those symbiotic algae are lost – unless the symbiosis can be re-established the reef dies. Other algae live in the fleshy 'lips' of giant clams, providing them with excess photosynthesis products in return for protection and, probably, micronutrients. One truly remarkable mollusc, the sea slug *Elysia*, consumes algae and then makes use of their chloroplasts which go on photosynthesising for a considerable time, relocated in the skin of the sea slug and turning it into a 'crawling green leaf' (Mujer et al. 1996, and for images of this remarkable animal visit: en.wikipedia.org/wiki/Elysia_chlorotica). If that does not hint at the ancient cellular origin of chloroplasts, nothing does.

Whole organisms enter into symbiosis with equal enthusiasm. Crabs use sea anemones or seaweed for protection. The decorator crab covers its pincers with tiny relatives of sea anemones, to take advantage of their stinging cells to kill small fish. Is this how tool-using began? Crocodiles employ birds as flying toothpicks to the benefit of bird and reptile. Oxpeckers live with zebra and cattle, eating external parasites. Bees pollinate flowers while collecting nectar and pollen. Ants farm fungi; the fungus gets propagated and the ants use it as a source of food. A good example is a small fish, a goby, that shares the burrow of the blind pistol shrimp, increasing the chances of survival of both. The goby uses its eyes and alerts the shrimp to impending attack by predators. The shrimp maintains contact with the goby by its antennae. When alerted, it snaps its claws with a sound that is compared to a pistol shot that scares or even damages the attacker. Human examples of beneficial associations include companion animals and the millions of bacteria in our guts. Even the vegetables that form part of our diet gain from their association with us. They get propagated in numbers that would not be possible in the wild. This list of mutual dependence goes on and on.

All of these examples, however, are superimposed on the most remarkable symbiosis of all, the great founding symbiosis that produced the modern cell. Sheep, cows, deer, crabs, sea anemones, seaweed, clams, algae, plants, fungi and humans are all constructs of the modern type of cell; each depends on the integrated activities of symbiotic modern organisms and the genomes of three or four ancient prokaryote ancestors. Our origins proclaim 'original symbiosis' rather than original sin!

In order to make the case for evolution, Charles Darwin buried his readers under a pile of unambiguous examples of evolution. In case you are not yet convinced, we have tried the same tactic here by providing you with yet more complex cases of symbiosis. Those who are now on the side of symbiosis may like to skip the next two pages. Never fear, however:

by comparison with Darwin's compendium, this is only a tiny list, but it justifies the central theme of this book, that life is a great universal cooperative enterprise.

The examples given here were gleaned from a short search of the literature. They testify to the fact that obligatory symbiosis is extremely common if not universal – although we think it to be universal. Here they are, in no particular order:

Bacteria living internally within insects provide essential nutrients in all species of insect so far studied (McCutcheon et al. 2009). As the Smithsonian has estimated that the number of species of insect is almost 1 billion, this is likely to produce a very large number of likely symbioses. As well as providing nutrients to insects, symbionts in insects can also have far-reaching indirect influences through insect- and plant-mediated effects at the community level. These include their impact on insect reproduction, on natural enemies of herbivores or on plant-associated microorganisms.

The pea aphid (*Acyrthosiphon pisum*) has an endosymbiont bacterium called *Buchnera*; its primary role is to synthesise essential amino acids that the aphid cannot acquire from plant sap (Wilson et al. 2010).

The tsetse fly *Glossina* has an endosymbiotic bacterium that is called, rather grandly, *Wigglesworthia,* a name that also commemorates a famous entomologist. *Wigglesworthia* synthesises vitamins that the tsetse fly cannot get from the blood it feeds on (Soumana et al. 2014). Without its endosymbiont, the tsetse fly could not survive – and the world would be free of the scourge of sleeping sickness.

Species of *Wolbachia* bacteria pop up everywhere (Bandi et al. 1999). Filarial nematodes (roundworms) are parasitic in humans, living in blood and lymph vessels. They live part of their lives as larvae in insects and are transmitted to humans when the insect bites them. They thrive in their intermediate insect hosts only because of an obligate endosymbiosis with *Wolbachia* spp. In many insect species, bacteria of the genus *Wolbachia* appear to play an important role in antiviral protection. The ovaries of the parasitoid wasp *Asobara* die if their *Wolbachia* bacterial symbionts are lacking. The wasps cannot then reproduce.

The mealy bug *Planococcus* lives on citrus trees. It harbors a bacterium, *Tremblaya princeps*, which, in turn, harbors a bacterium, *Moranella endobia* – rather like Russian dolls (McCutcheon and von Dohlen 2011)!

Obligate wood-digesting endosymbiotic protozoa live in the intestines of termites. The symbiosis is obligate in that neither the protozoa nor the insect can survive without the other (Sutherland 1933). We chose to cite this paper because it describes the discovery of Margulis's (1970) pin-up symbiont, the wood-eating *Myxotricha*, a fivefold symbiont that is itself symbiotic within a termite.

Riftia is a marine worm that lives its strange life in close proximity to black smokers. *Riftia* lacks a gut and so relies for nutrition on endosymbiotic bacteria that can deal with this extreme environment (Bandi et al. 1999).

Symbiodinium is one of the dinoflagellates, ancient cells of the blue-green algal sort. It is a photosynthesiser and it is found as an endosymbiont in corals and molluscs, providing them with photosynthetic products in return for shelter and other nutrients (Fitt 1984).

Symbiotic algae on the egg jelly of salamander eggs produce the oxygen necessary for the survival of the spotted salamander embryos (Kerney 2011). Newborn of the squid *Euprymna scolopes* lack a light organ; it is only developed in cooperation with luminescent bacteria (*Vibrio fischeri*) in their skin (Ruby and Lee 1998).

The development of the immune and the digestive systems in mice raised *without* gut endosymbionts cannot be completed. All known mammals, including humans, have gut symbionts that contribute to their wellbeing (Round and Mazmanian 2009). In the zebra fish, microbes regulate the normal proliferation of the intestinal stem cells. Without these microbes, the intestinal epithelium has fewer cells, and it lacks goblet cells altogether, so the fish cannot survive unless it can recruit its symbionts (Kanther and Rawls 2010).

In plants, there are fungi that live out most of their life cycle in plant tissue and improve immunity in the plant (Wani et al. 2015). They also discourage herbivores – perhaps by tasting nasty.

We have, however, left the best to last. In recent years much work has been devoted to 'common mycorrhizal networks'. These are networks created by fungi living on dead plant material, such as leaf litter and humus in the forest floor (Tlaika et al. 2008). The fungi connect plants to one another by fungal tubes called hyphae. The networks can transport water and soluble compounds containing carbon, phosphorus and nitrogen from one plant to another. They can also transport molecular 'alarm' and 'defence' signals. Some networks are not fussy and connect up all the plants in their neighbourhood, while others concentrate on a single group of related species. This net runs through woodlands and forests like a closed computer system, ensuring all the trees are connected. Not only this, but surplus photosynthetic products from the leaves in the forest canopy flow down the trees to the roots and are distributed by the fungal network to the shaded plants below. It is an excellent example of energy flow from the 'source' (the sun on the canopy) to the 'sink', (the denizens of the forest floor). It is an ancient networked communication system.

We rest our case. It is very doubtful if there is a single animal or plant that does not have at least one symbiont.

But the story doesn't end there. There is the problem of parasitism. Is a parasite merely an organism on the way to negotiating a beneficial symbiotic relationship with its host or is it something else? The world is full of parasites. One definition of a parasite could be:

> an organism whose survival depends on the genome of another organism, to the latter's disadvantage.

Yet again, a reductionist definition wants to restrict parasitism to a single relationship, the genetic one. It ignores within-species parasitism, the most human-referenced examples of this being slavery and cannibalism.

All known viruses are obligate parasites by this definition, as they move in on cells and take over their genetic mechanisms to make more viruses, after which the cells die. Some, very few, bacteria among an uncountable number of different sorts are parasites and cause disease in humans.

Arthropods, belonging to the great phylum containing crustaceans, insects and spiders, are particularly innovative when it comes to the development of parasitic forms. One eats out the tongue of a fish and simply replaces it with its own body and remains in the mouth, acting like the original tongue (Eman et al. 2014). It is an isopod, belonging to the same order as the slaters and woodlice in your back garden. Is this symbiosis or parasitism? *Sacculina* is a parasite of crabs. It belongs to the barnacle family – yes, they are arthropods! – and burrows into crabs, castrates them and lives happily ever after (Lafferty and Kuris 2009). Pea crabs are tiny crabs that live within the shells of oysters and mussels and rely on them for food and shelter.

It is obvious that mutually beneficial biological relationships are not created overnight. They take generations to establish. The pistol shrimp and the guppy did not negotiate accommodation – it just happens that guppies living in pistol shrimp holes did better than their neighbours and so did the shrimp. The pistol shrimp had to learn that the turbulence of the water, as the guppy fled into its hole meant danger and to react with its deafening and disorienting 'pistol shot'.

Parasitism is so common that we also use the word freely in describing aspects of human relationships, where one person 'sponges' on or is otherwise dependent on another. (This, by the way, is a terrible slander on an inoffensive group of animals that do little but sit on the seabed and filter sea water!)

Definitions to scientists are thus rather like words to Lewis Carroll's Humpty Dumpty:

> 'When I use a word,' Humpty Dumpty said in rather a scornful tone, 'it means just what I choose it to mean – neither more nor less.'

Every organism so far studied has at least one parasite; it is a habit of life that is so common that one can confidently claim that the number of parasites on this planet exceeds the number of the free-living organisms – by at least one! Jonathan Swift understood this as far back as the eighteenth century:

> So, naturalists observe, a flea
> Hath smaller fleas that on him prey;
> And these have smaller still to bite 'em;
> And so proceed ad infinitum.

Now for symbiosis and parasitism in humans. The way the relationship between two people commonly develops provides an analogy to clarify the process.

The beginning of a human being follows the obvious symbiosis between ova and sperm: they both die if they do not meet. The developing foetus is totally dependent on the mother, a very intimate relationship that some have argued is essentially parasitic. Once born, the child must be raised, and this requires another individual or individuals or they do not survive. Children have often been raised by their siblings or by their grandparents. Sometimes they have even been 'raised' by wolves or other animals, although the idealised life of Mowgli in Kipling's *Jungle Book* is a very long way from the sordid truth. According to Mary-Ann Ochota, these children are unwanted and usually the victims of extreme neglect (www.youtube.com/watch?v=3n7ZtATu0cU).

The fact remains that it occurs and provides an example of human–feral animal interaction that is at least mutually tolerant. Normal development in children, however, requires the establishment of a relationship with at least one other human, male or female. Speech and empathy do not develop without it. The parent ('host'!) is genetically programmed for an emotional bond with the newborn ('parasite'!), accentuated in mammals by the connection of breastfeeding.

As the child grows, what are the characteristics of the symbiotic relationship between it and other humans? Is it equal and caring or is it asymmetric, with one taking advantage of the other? Sadly, when a child is not accepted as part of a relationship, it can be rejected as if it were indeed a parasite. Finally comes the establishment of a fresh relationship between two adults, and cohabitation follows. Cohabitation may turn out to be a great success – or sadly, may turn out to be toxic for one partner who is really in the role of prey.

This description of the human condition is not really analogy at all. Peel away the human associations and you have a perfect description of the establishment of a relationship that can be symbiotic but may be exploitative. Parasitic relationships between two organisms are asymmetric and are therefore in the toxic range. In extreme form in humans, this could be displayed as domestic violence.

Here are three instances of parasitism. The stages discussed above are readily identifiable.

Naegleria is a free-living amoeba that lives in soil and freshwater. Occasionally, in warm countries, the amoebae form a bloom in water holes, and someone who jumps in may force them up their nose (contact). The amoebae can move anywhere in the body and most won't survive.

But some may migrate along the olfactory nerve (cohabitation) into the brain (John 1982). There, they proliferate and cause lesions that are frequently fatal (extremely toxic).

Humans gain many of their parasites from the animals with which they live in close association, not necessarily by choice. These pathogens may have preference for animal hosts, but occasionally they infect humans – perhaps the unfortunate person is immunocompromised in some way. That makes human-to-human infection possible and that in turn gives the invader a chance to become adapted to humans. It now has the opportunity to infect more and more people and the animal source may later be cut out of the cycle. Thus, while rabies is only contracted from rabid animals, Ebola and dengue viruses continue to be recruited from an animal source, as well being transferred between humans, while AIDS is transmitted predominantly between humans.

Civilisation offers some unexpected opportunities for establishing interspecific relationships. Foxes have their own breed of tapeworm. Riding to hounds, a favourite pastime of the nobility, provided the opportunity for a parasite to extend its range. Gradually the fox tapeworm was transmitted to the pack of hounds, and from the hounds to the dog handlers and members of the hunt. Oscar Wilde, who wittily described foxhunting 'as the unspeakable in pursuit of the uneatable' would have been pleased to know that the 'uneatable' member of the hunting relationship got its own back on the 'unspeakables' by bequeathing them a unique and aristocratic type of parasitic fellow traveller.

The evolution of the malaria parasite is yet one more example that demonstrates the readiness of different genomes to cooperate, one within another, which characterises so much life on Earth (Okamoto and McFadden 2008). An ancient cell, a blue-green alga, was first engulfed by a modern cell but was not digested and formed a symbiotic relationship with its host cell. The result of this primary endosymbiosis was a photosynthetic alga, of the modern type. A descendant of this alga was then itself engulfed by a non-photosynthetic modern cell and another symbiotic relationship was established. This complex arrangement eventually became the malarial parasite transmitted by mosquitoes, causing disease in people and animals. This suggests that there are many forms of modern cells containing different combinations of inclusions, which in turn indicates that the type of symbiotic event that produced them was common.

Is parasitism simply an association between two organisms that have not yet learned to live in mutual harmony? Mutual harmony is the level of true symbiosis where both or all of the partners benefit. Perhaps this is the level to which all parasitic associations eventually aspire, but as we have caught them in mid-development, as it were, we have a tendency to think that what we see are end points whereas they are only end points thus far.

Which brings us to the question: what is an individual person? Every one of us is a cluster of interacting genomes all contributing to our mutual survival. Our good health depends on hundreds of different species of ancient cells living on and in us. These constitute the human biome and there are many papers in the literature ascribing a variety of human disorders to perturbations of this biome. In 2016, the US government announced a $500-million project to explore this 'inner space'.

The fact that we depend on our intestinal flora and fauna for a healthy life explains why living in a totally clean environment might not be as healthy as those who sell disinfectants would like us to think. As an example, the sale of disinfectants that kill more than 90 per cent of known bacteria in a porcelain toilet bowl is, quite apart from being a cynical money grab, no more effective than flushing with water. Our world *ought* to be a bit dirty, so that we can continually recruit our beneficent allies and hone our capacity to repel boarders. You, and me, and the whole living world of plants, animals and fungi, each one of us is a holobiont, a community of organisms. There is more opportunity for natural selection to work with than Darwin ever dreamed of.

The biological world is a much more wonderful place to be today than it was 60 years ago when the authors both 'read' biology at a university. Modern philosophers of science now allow us to ascribe a form of purpose to life and, of course, as we have always suspected, living systems do have purpose. Darwin certainly thought so. That purpose is maintaining the self-regulating system, staying alive and staying alive long enough to reproduce. We are now allowed to believe that an organism can exhibit intelligent behaviour without being conscious and we can allow sentience at various levels to a wide range of animals – mammals, birds, octopuses – but dog and horse lovers have always known this. Natural selection is still with us but nature is not necessarily red in tooth and claw. It is far more often cooperative.

A study of evolution shows that complexity increases with time. Organisms do not become more complex by inventing brand-new ways of doing things; instead, evolution relies on processes that have been successful in the past and repeats them at higher levels of organisation. A consistent pattern appears. Evolutionary novelty arises after a period of collaborative experiment, the outcome of which is subject to Darwinian natural selection that rejects or accepts the innovation.

A metaorganism (like you and me, and sharks and tomato plants and mushrooms – think of all of us as primary bionts) is not a standalone entity. It provides a series of perfect environments for a whole range of other organisms. We will ignore for the moment such mega-parasites as worms and fungi and consider only ancient lodgers such as bacteria and other single-celled organisms. It is now thought that we humans represent housing for the accommodation of hundreds of different species of age-old symbionts. They live on the skin, in the alimentary tract, nasal cavities, reproductive system – in fact, almost anywhere. One casualty of these ideas is the thought that one might recover the ancient DNA from a frozen mammoth, say, and use it to recreate the extinct species. Without knowledge of its fellow travellers, however, the experimenters may be doomed to disappointment. If they haven't done so already, the researchers would do well to return to their frozen specimen in Siberia and see what they can resurrect from its intestine!

There is today great interest in humans as holobionts. The National Institutes of Health Human Microbiome project is concerned with the study of all the organisms that are to be found in the complex ecosystem that is the human body and make up to 3 per cent of our bodyweight. These microbes are generally not harmful to us, in fact they are essential for maintaining health. They produce some vitamins that we do not have the genes to make, break down our food to extract nutrients we need to survive, teach our immune systems how to recognise dangerous invaders and even produce helpful anti-inflammatory compounds that fight off other disease-causing microbes.

A vanishingly small proportion of microbes is responsible for human disease. The majority of them, such as *Yersinia pestis*, which causes plague, are just opportunistic freeloaders, a by-product of whose activity is the death of the victim, the holobiont. Sometimes, an individual can come to an equilibrium with a disease organism. 'Typhoid Mary' was such a person; a cook, she carried *Salmonella typhi* without harm to herself and left a trail of disease and death behind her.

Changes in the composition of the gut biome correlate with certain disease conditions. New techniques, such as faecal transplants, are found to be effective in treating them. Several autoimmune diseases are improved by infecting the patient with intestinal worms. It is as if their immune systems have been misdirecting their well-meant activity and the worms give them a more appropriate target to aim at. It is also possible that inhabitants of our intestines have a greater say than we ever thought in our peace of mind. Our gut symbionts can produce almost every neurotransmitter that keeps our brains in good working order. It may be that anxiety and depression can be treated by regulating their output.

Everywhere you look there are cycles, and cycles of cycles. Humans and human social systems are part of, not separate from, this huge and intricate system. Dynasties rise and fall, families go from rags to riches and back again. The cyclical nature of life has not been lost on historians. Spengler, writing during World War I, in *The Decline of the West* (1932 reprint), made the intellectual leap of considering human society to be an ecological system, or even an organism, in that it experiences periods of growth, maturity and decline. Toynbee, in *A Study of History* (1972), saw the parallel with evolutionary innovation that necessarily occurs at the boundaries of two or more conflicting ecosystems, and suggested that new civilisations arose when faced with physical or social challenges provided by those already present. If societies fail to meet these challenges, they have a tendency to fall apart.

The long history of humanity is the history of the biosphere. Interdependence is one major strand of the golden thread that connects the biosphere. Along with energy flows and genetic inheritance, and for humans, language and social customs, the thread becomes part of Gaia's self-maintaining systems that make our world.

9

WALLING IN AND WALLING OUT

In which we recognise the influence of boundaries in encouraging cooperation and innovation.

Cooperative evolution links all living things, making connections that become the planet-wide system of Gaia. The interfaces between the cell and the environment, between one cell and another, between species, between ecosystems, all facilitate evolution by presenting challenges. Over and over again, what have appeared to be boundaries have turned into launching platforms for new forms of life.

Darwin was clearly aware that the origin of life can only be inferred from our observations of today's Earth. He saw the necessity for concentrating life's inorganic precursors within a small container, like a cell membrane, to promote the formation of connections between them. In a letter to a great friend, Professor Joseph Hooker, dated 1 February 1871, he imagined a 'warm little pond' as a device for creating a place where all the conditions necessary for the emergence of life could be brought together. Its edges, its surface and its bottom provided its boundaries with earth and air. The water provided the solvent for the chemicals of life to connect and react. The pond was warm, to provide the energy necessary for the reactions of emerging life to take place. Darwin even suggested the types of chemicals – 'all sorts of ammonia and phosphoric salts' – that could accumulate in the pond in sufficient concentration to generate the precursors of life, such as 'a protein compound ... ready to undergo still more complex changes'. This is a vivid metaphor to show how a bounded system might create and sustain life as well as exclude non-life. From the beginning, boundaries have been as important as the contents they held. A single cell is a microcosm of that warm little pond.

Modern thinking has allowed Darwin's warm little pond to flow away. It was a remarkable metaphor for his time and, like so much of his work, it is not inconsistent with modern thinking. It is, however, now thought that the environment in which these first steps occurred was something like that created by deep-sea geothermal vents, black smokers. It seems likely that life evolved at high temperatures in a watery milieu that contained the right set of chemicals. The edges of the pond have been replaced by boundaries created by temperature and concentration gradients as the chemically rich hot water from the vents diffused into its cooler surroundings, creating an intermediate 'Goldilocks' zone where everything was 'just right' for the beginning of life.

Much thought is also being given to the idea that these early stages might have taken place on Mars (see, for example: www.space.com/22577-earth-life-from-mars-theory.html).

Mars had the right conditions about 4 billion years ago. Might the precursors of life have been transferred to Earth when meteorite impact or enormous volcanic eruption hurled out rocks that reached escape velocity and eventually fell to Earth? If there is any merit in the suggestion that the pre-cellular seeds of life originated on Mars, then crossing the boundary created by 225 million kilometres of interplanetary space was a stupendous event in life's history. It set the scene for billions of years more of negotiating boundaries. From the simplest to the most complex, every form of life exists because of a boundary that separates it from the non-living and distinguishes it as an individual.

According to the *Oxford English Dictionary*, a boundary is 'a thing which serves to mark the limit of something: the limit itself, a dividing line'. Robert Frost, in his poem *Mending Walls*, saw a boundary in terms of its function;

> Before I built a wall I'd ask to know
> What I was walling in or walling out

This is good advice. Biological walls, however, go far beyond this simple, poetic dichotomy; they are dynamic. It is possible to tell the whole story of organic evolution through boundaries, from black smokers to human skin.

Boundaries of non-living things may be inert and impermeable, important features that have led to people creating containers of all kinds, from biblical wineskins to leather handbags. When a boundary is merely a wall, nothing can get past it, neither in nor out, in which case they are merely containers in which the contents are preserved. Connection and collaboration with the outside world cannot take place. Nothing much can happen to anything within except its eventual decay. That thermodynamic tax, entropy (the tendency to disorder discussed in an earlier chapter), ensures this outcome. Even a vintage wine held in a tightly sealed bottle eventually becomes undrinkable.

Here is a thought experiment. Fill a tin can with blue-coloured fluid and set it upright in a bowl of water. Nothing gets in or out. The wall of the can is an impermeable membrane, a wall. Remove the can, empty it, perforate it with pinholes, add more blue liquid, and replace it in the bowl of water. You will immediately see coloured water coming out through the holes and mixing with the water in the bowl. This process will continue until the intensity of the colour is the same, inside and out. It is then in equilibrium. Nothing much else can happen without some other input from an external source. The perforated can is like a freely permeable membrane. In order for rather more exciting molecular events to take place, movement through the tin wall must somehow be controlled. In biological systems, membranes are not simple inert barriers with holes in them. They are dynamic structures, composed of fat and protein molecules; they are fluid membranes, in constant motion.

Biological membranes often have the property of semipermeability. A semipermeable membrane is a membrane with pores, but whose pores are of a size that will allow small molecules to pass through, in or out, but not larger ones. There is nothing magic about semipermeable membranes, as they can be readily created in the laboratory. Better still, flattened tubes of flexible, semipermeable, artificial membrane for experimental purposes can be purchased from a chemical supply house!

Here is another thought experiment. Cut off a length of the tube, knot one end and half fill it with a concentrated solution of household sugar in water. Squeeze out the remaining air and knot the other end to make a sort of half-filled sausage. Immerse it in a bowl of water and watch what happens.

The molecules of sugar are much bigger than those of water and they cannot pass through the pores of our semipermeable membrane. Gradually the tube begins to swell as water enters it. This process is called osmosis. No sugar molecules get out of the tube. The much smaller molecules of water, always in motion, pass through the wall of the tube. In fact, they can go in *and* out freely – except when their exit is blocked by a sugar molecule. As long as the rate of entry of water molecules exceeds the rate of departure, the spare space in the tube will start to fill up with water. Eventually, the system will come into equilibrium when the back-pressure exerted by the gradually filling tube is equal to the pressure of the water molecules bombarding the outside of the tube trying to get in. The pressure can be measured and is called the osmotic pressure.

The outer wall of a cell is a semipermeable membrane. Water always tends to get in. Because of this, when the cells are immersed in water, it enters the cell and must be disposed of to maintain the correct concentrations of important cellular contents. An amoeba, which lives its life in fresh water, has to work hard to bail it out through its contractile vacuoles.

The creation of a semipermeable membrane, enclosing all the cell metabolites, was one of the defining points in the evolution of the cell. Cell membranes, however, are not just semipermeable; they also have numerous pores of different shapes and structures. In the simplest case, think of a pore in a membrane as a hole controlled by a five-barred gate. Water molecules, which can pass between the bars of the gate, go in and out freely, while those of larger size are stopped. This is why amoebae have contractile vacuoles that continually pump out water. Energy has to be expended to do this work and any small molecules of soluble waste products can be eliminated as well.

A selectively permeable boundary is one that contains a number of 'gates', variously hung. Some swing both ways, letting molecules that are smaller than the gate go freely in and out. Other 'gates' only open inwards, letting molecules from the outside in; others only open outwards, letting molecules from the inside out. A cell has many gates under its control, gates that can exist in either of two conditions: locked or unlocked. Desirable molecules – from the point of view of the cell – that are outside the cell hold 'keys' to the locks. The lock is precisely shaped so that these molecules can easily attach to it and gain entry. On the inside, molecules that need to get out also have keys to their appropriate gates. The cell is thus a device for maintaining inside itself a tiny internal environment that is totally under its control.

The concept of a dynamic boundary extends well beyond the cell membrane. A multicellular organism like a fish is covered by a membrane made up of different sorts of cells that we call skin. The continued integrity of the skin is essential for maintaining high level of organisation of the systems contained within. In fact, the immortal words of Alan Sherman, 'skin's the thing that if you got it outside, it helps keep your insides in' apply both to you and an amoeba.

A cell membrane (or a skin) is an interface with the environment, and regulates all traffic, in and out. In doing so, it has to distinguish between 'good' molecules and 'bad'. The cell is right to be suspicious, because the 'bad' molecule might poison an important internal process.

In a multicellular organism, a metaphysical boundary (a sense of 'self') is created by the continued interactions involved in overcoming potential invaders. Distinguishing 'self' from 'non-self' is so important that it creates a heavy selection pressure in favour of the evolution of immune systems where the recognition of 'self' leads to responses that protect it from pests that are 'not-self'. Its finest expression is to be found in the exquisite immune systems of mammals and birds, because their body temperature (36–40°C) is ideal for bacterial replication (Pradeu 2012).

This metaphor of the gate applies equally well to, say, a house, a village, a walled town, a city and even a country. The concepts of self and not-self govern human behaviour, even at a societal level. From school football teams to national pride, ideas of, and distinctions between, 'us' and 'them' are strong motivators for preserving boundaries. Flocks of birds, shoals of fish, herds of buffalo and teams of people: all are affirmations of 'us' and the members share the risks presented by others that are 'not-us' by acting together, to the advantage of the whole population.

In Darwinian evolution, the physical and physiological demarcation of boundaries has always acted as a potent generator of novelty, because selection pressures are greatest where two or more boundaries meet.

If a species is to 'cross over', to adapt to the new circumstance, it must display sufficient variation to provide individuals with the requisite adaptability to allow them to flourish within the new environment. What it already possesses was defined by the old environment, yet it must be flexible for use in the new one. Thus, the swim bladder, a flotation organ found in bony fish, provides the possibility for air-breathing lungs in a lungfish; a pectoral fin becomes, variously, a wing of a pterosaur, a bird or a bat or, full circle, the fin of a whale.

The boundary crossings from one ecosystem to another are physical or geographical. To an organism, the environment is effectively a complex system of edges, a jigsaw puzzle, each piece carrying a different suite of attributes and a new roll call of foreign organisms. It is the same wherever you look, even in three dimensions. In the sea, plankton drifts with the current and undertakes daily migrations up and down in the water column. One boundary for plankton is the interface between water of different temperatures; another is the interface between water and land or ice. For the blue whale that feeds on plankton, survival depends on responding to this complex environment.

Many types of boundaries exist. They may be physiological. The transition zone between sea and freshwater is a barrier to the many organisms that lack either the physiological equipment to deal with the osmotic stress exerted by fresh water or, going the other way, with the salt loading imposed by the marine environment. An estuary offers a huge challenge; the mouths of rivers are populated by species that have learned to survive in waters where the salt content is low or variable. Further upstream, the inhabitants have learned to cope with freshwater and now cannot survive in saltwater. Humans are able to cross these potent environmental barriers only by means of technical solutions – diving bells, wet suits and boats.

Humans have their own barriers. Physical characteristics can act as social boundaries. Language, skin colour, eye shape and food customs have in the past isolated communities as if they existed on islands. With time, the differences are sufficiently emphasised to create whole nations. Think of China, Europe and India. Social boundaries in turn can reinforce geographical boundaries.

Behavioural boundaries are common. In a widely spread population of birds, say, differences in sexual behaviour may become entrenched at the extremes of their distribution. If birds from the extremities are reunited, their behavioural 'accents' may restrict or even prohibit breeding. Such birds are known as 'ring species'. Greenish warblers form a species ring around the Himalayas. They 'set out' from their supposed point of origin in Nepal, form a chain of breeding around the Tibetan Plateau and meet up again in Siberia. By this time, their geographical displacement has led to the creation of a new breeding barrier. Nevertheless, it is still feasible, if unlikely, for a mutated gene to travel the whole circumference of the ring. (We are grateful to the late Professor Colin Groves, who drew our attention to the greenish warbler. Apparently the first example of a ring species we chose, a European seagull, is no longer considered to be one.)

If this effect of separation is difficult to accept, consider the more recent history of the English language. In 1607 the first successful English colony was established in Virginia, in what would become the USA. As the colony expanded, pioneers moved inland and settled in the Appalachians. Here they established new settlements and had little commerce with the mainstream colony. In this relative isolation, their dialect maintained its strong Elizabethan accent and vocabulary so that it has become almost incomprehensible to native English speakers from the south of England in the twentieth century (West Virginia Archives and History: www.wvculture.org/history/journal_wvh/wvh30-2.html).

The Romani (previously referred to in English as 'Gypsies') are notable among human societies for their nomadic lifestyles. Originally a wandering, low-caste Indian tribe, they were transported to Persia in the fifth century and their subsequent diaspora has led them to many countries of the Middle East and Europe. In their journeying they adopted many of the social *mores* of the lands in which they travelled, at the same time retaining a version of their own language for use among themselves. People at the opposite ends of the migration would have trouble communicating without their own specifically conserved language (see: en.wikipedia.org/wiki/Romani_people).

On land, geology and geography combine to create biological boundaries that control the distribution of plants and animals. The occurrence of barrier mountain ranges and of different rock types, varying soil structure and components, availability of water, rainfall, temperature, streams and rivers, all create an intricate mosaic of environmental types.

Human activity contributes to this complexity by adding the selection pressures of hunting and gathering, agriculture and animal husbandry, and the construction of towns and cities and their infrastructure. Neolithic human societies created the first cities in Mesopotamia 8,000 years ago. City walls were attempts to regulate and keep people safe within their newly established environments. Within the cities, humans set up social structures that were defined by the metaphysical boundaries of 'us' and 'not-us'. All this created, and continues to create, new biological niches, opportunities for life forms old and new – such as urban foxes and European magpies in London and the ever-opportunistic rats and mice. But think also of the opportunities offered to pests and parasites by humans living close to one another in villages and in close contact with animals – plague, tuberculosis, mumps, measles, smallpox, coronavirus and HIV/AIDS. So far we have risen to the challenges that they presented us with, but there is always something poised to have its day at our expense, such as Ebola virus. Crossing boundaries is an important property of life and it happens continually.

Two personal examples illustrate the complexity of boundaries, how easy they are to create and, paradoxically, how difficult they may be to detect. One of us (Bryant 1983) worked for many years on the biochemistry of parasites. One of these projects involved a nematode (roundworm) that is parasitic in rats. This particular nematode is a much-used experimental organism, but intense study – very reductionist! – showed that there were differences in the biochemical profiles of worms maintained by his research group in Australia and those of a Canadian group. This was odd, because both cultivars were derived originally from the same source. After much puzzling and comparing of notes, the source of the variation was identified. The cultivars were kept under identical conditions – except that in Australia, the research group collected rat faeces containing parasite eggs in the morning while in Canada, they found it more convenient to collect them in the evening. Rats are diurnal and their physiology varies during the day. This results in 'morning eggs' experiencing a different physiological climate from 'evening eggs', which then was reflected in changes in the biochemistry of the hatched worms. In retrospect, this difference might be attributable to epigenesis – another chance of fame and fortune missed!

The second example concerns human groups. Valerie Brown (Brown and Harris 2014) once worked with local councils within the single province of Western Sydney, trying to establish a program of coordinated environmental management across the whole area. Once staff from the different councils came together it became apparent that the differences between them far outweighed their desire to collaborate. The differences flowed from different environmental philosophies, creating subtly different environments in their areas of jurisdiction. This exerted different selection pressures on their human residents (and many cohabiting species). For instance, the category 'open space' was defined variously as parkland, wilderness areas, car parks, sporting fields, community gardens and even light industrial areas! As well as these differences, councils might have a closed hierarchical management style, or a long tradition of community partnership. This wide range of potential difference was not only apparent in physical differences in their territories, but also in different ways of life within the same area.

The fact that people are able to teach their offspring means children learn from preceding generations. In addition, humans preserve their knowledge in books, parchments, papyri, vellum, clay tablets, metalwork, stone tools, cave walls and even fossils. The barrier created by the inability to read another civilisation's scripts is illustrated by the Rosetta Stone. It was only possible to decipher it because three different scripts were used in turn to describe the same events. The Rosetta Stone is a little more than 2,000 years old. The human insights of 200,000 years, from stone tools to the present day, are available to those with the skills to read the artefacts.

Each one of the means of communication mentioned in the previous paragraph represents a lesser or a greater barrier encountered and overcome. The first stone tools solved fundamental problems of survival, making fire (flints), keeping warm (prepared animal skins) and providing access to high-energy food (butchering). The cohesiveness of family groups and their interaction with others enabled the new knowledge to diffuse laterally and also to flow down the generations and – uniquely – to persist, to *accumulate*. As each barrier was crossed, the emergent consequences were enormous. The stock market is implicit in the first friendly barter of a stone axe for an animal skin, the Industrial Revolution in the first successful smelting of iron ore.

To state yet again the basic proposition of this book, the story of life, from the first appearance of ancient cells, is one of cooperation. It is a story of mutual encouragement and mutual aid. The part of it that is the animal kingdom is a very small part of the whole. Across the whole biota, cooperation and competition have been, and still are, occurring simultaneously. Among the oldest forms of cooperation between organisms is a process – sexual reproduction – that started at least 3 billion years ago and is with us today. It is not just an interesting biological sidelight but an important evolutionary engine.

Negotiating a way of life between two or more ancient cells to create the modern cell billions of years ago is responsible for a phenomenon of such biological imperative that it still obsesses humanity to this day. That is sex; its physical manifestations and its psychological effects drive so much of human creativity. Biologists have had difficulty explaining why such an apparently risky phenomenon as sexual reproduction should be maintained in nearly all organisms except possibly an immortal jellyfish or two. The usual explanation is that sexual recombination increases variation and provides the raw material for evolution.

Think, however, of the number of boundaries that the reproductive process in humans has to bridge, without harming the evolutionary desirable outcome in the process. They are, first, the problems of human relationships; next the lottery of sperm and ovum in fertilisation; then the partitioning of the sex chromosomes X and Y. Many genetic disorders occur at this stage, but the risks are the price of creating genetic variation, the raw material for future survival.

So, why so much sex? Asexual reproduction is certainly not out of the question. It is common in plants and in many animals. Stem cell research has already given us the ability to clone animals. A compelling new hypothesis derives from the intrinsic properties of the ancient cells that make up the modern ones (Havird et al. 2015). In animals, the rate of change of mitochondrial DNA is about 10 times that of nuclear DNA. Cell division in the nucleus during sexual reproduction creates considerable DNA variation. It is argued that this is necessary because it provides sufficient novelty to allow the nucleus to keep up with mitochondrial variation and thus maintain the exquisite regulatory balance that must exist between the two.

Geneticists and genealogists both exploit the high rate of variation of mitochondrial DNA to determine relatedness. It is relatively small – in humans, about 16,000 base pairs code for 37 genes – easily identified and sequenced, so differences between two samples can be readily identified. There is, however, an additional reason why it is so useful. The nuclear DNA of the human tadpole-like sperm cell is all in its bulbous head.

The mid-piece of the sperm is tightly packed with mitochondria that supply the sperm with the energy necessary for the arduous voyage to the egg. Fertilisation occurs when the successful sperm contacts the egg. At the outer membrane of the egg, the sperm sheds it tail *plus its mitochondrial mid-piece*. Only the head containing nuclear DNA enters the egg to trigger development, while the mid-piece is discarded, like the exhausted booster of an interplanetary rocket ship. All the mitochondria in the resulting developing embryo are therefore derived from the maternal ones that are already present in the egg.

In plants the rate of mitochondrial variation is much less, though still greater than in the nucleus. It is for this reason, perhaps, that plants are more able to undertake continuous asexual reproduction. However, with time, mitochondrial variation will still become too great, so that sexual reproduction once again becomes necessary to allow the nuclear DNA to catch up.

One possible reason why the rate of change of mitochondrial DNA is greater than that in the nuclear DNA is because oxygen is dangerous stuff (Baker and Orlandi 1995). Mitochondria have to interact intimately with oxygen as it is used in energy metabolism. In making ATP (adenosine triphosphate), the 'energy currency' of the cell, they transfer electrons to oxygen to yield a molecule of water. In this process, highly reactive intermediate oxygen products that react with almost anything, including DNA, are produced. Damage to DNA is a constant possibility that must be avoided and there are protective molecules ready to scavenge the dangerous oxygen radicals. Even so, some of the highly reactive oxygen compounds do escape to do damage to important molecules and subcellular structures. Plants have to deal with the perils of oxygen in two systems; chloroplasts as well as mitochondria. Chloroplasts once were free-living anaerobic photosynthesisers *producing* oxygen as an end product of photosynthesis, and therefore have a much longer history of dealing with toxic oxygen than mitochondria. Perhaps they are better at it. They too have their antioxidants to deal with reactive oxygen. In any event, modern cells successfully crossed this barrier to symbiosis and the rest is all about you.

Those first great cooperative events that led to the modern plant and animal cells set the scene for the rest of evolution. The next great innovation occurred when two modern cells – two separate but identical cells – started to work together to enhance their mutual survival. This was the first step in opening up a new range of environments that could be occupied by larger organisms, made up of large numbers of modern cells. The exploration of the advantages of larger size was at last possible.

If the regulatory adaptations necessary for integrating two dissimilar modern cells were too difficult to acquire (after all, that means integrating six or eight different genomes of separate ancient origin), those involved in integrating the activities of two similar genomes were resoundingly successful, and happened at least three times, leading to the kingdoms of animals, plants and fungi. The evidence for this is all around us, and this 'major transition' appears not have been so difficult as you might imagine. 'Snowflake yeast' is a variety of the usually single-celled yeast, *Saccharomyces cerevisiae*. A single mutation enables the cells to stick together to form the 'snowflakes' and thus open up the evolutionary door to multicellular organisms (Ratcliff et al. 2015). This also is the case among algae, where a single gene allows the daughter cells of the alga *Gonium pectorale* to stick together (Hanschen et al. 2016). If only one gene alteration is necessary to create multicellular plants and animals, it becomes a racing certainty.

Once modern cells came into existence, the door opened for the evolution of another layer of symbiosis. Once two or more modern cells successfully collaborated in survival, the scene was set for an explosive adaptive radiation into ecological niches that were not available to single-celled organisms. As the aggregations of cells grew larger, problems of communication within the cluster began to exert their effects. The problems were solved in the first instance by specialisation; the different tasks of survival were allocated to cells that were equipped to deal with them. An early stage of this level of organisation can be seen in modern sponges that have only a few cell types.

Sponge cells, however, have not yet completely lost their individuality. A well-known experiment involves passing the sponge through a fine mesh cloth to separate the cells. The cells survive the experience and return temporarily to an unspecialised condition. With time, they gradually reassemble into one or more smaller sponges. This apparent simplicity may, however, be illusory as sponges have an evolutionary pedigree as long as any of us. One sponge has recently been shown to possess nearly as many genes as humans, about 18,000, suggesting a rather more complex ancestry than originally thought (Mann 2010).

The next level of organisation is found in *Trichoplax,* the only living member of its own phylum, Placozoa (see Srivastava et al. 2008). This tiny animal is barely a millimetre in diameter and has been found only in aquaria. In *Trichoplax,* cells of similar function are grouped together, into two tissues, a sheet of skin-like cells on its top surface and motile and secretory cells underneath. It moves across surfaces and grazes on any ancient cells that it encounters. It has a small genome of 13,000 genes, but quite a few of these are similar to those found in higher organisms.

With greater complexity and size comes a need for greater organisation. In response, tissues of varying sorts that contribute to a single function are gathered into organs. As an example, consider the liver, a very complex organ indeed, with many different, permanently differentiated cells serving the range of functions that all contribute to the property of 'liver-ness'. Plants, being modular, never invested so heavily in the specialised organ concept. Cutting off a branch does not kill a tree; removing the liver from a mammal is always fatal.

The way that evolutionary complexity has occurred emphasises the inherent capacity of living systems to cooperate, both with each other and then at a higher ecological level. In each case, the evolutionary process works with what has gone before to create novelty. If ancient cells can work together to produce a viable unit, why not try aggregating modern cells? If some cells are all doing the same thing, why not clump them together into tissues? If tissues are combining to serve a single complex function, why not gather them into organs? Perhaps even whole animals and plants can be clustered to provide additional novelty to make an ecosystem. Perhaps ecosystems can be clustered to make … Gaia?

And that, of course is what has happened. There are many examples of animals and plants that have adopted a colonial or social lifestyle. *Dictyostelium,* the little slime mould that we have already met, has been used as an experimental model for development and differentiation (Durston 2013). It is a member of a very successful group of protists that have adopted a cooperative lifestyle. JT Bonner, a world authority on slime moulds, describes them as follows (Bonner 2010):

> No more than a bag of amoebae encased in a thin slime sheath, yet they manage to have various behaviours that are equal to those of animals who possess muscles and nerves with ganglia – that is, simple brains.

So, becoming a social organism works for the slime mould; then what about higher animals? And what about humans? Well, the evolutionary muse might have remarked to herself, what worked once will work again.

'Go to the ant, thou sluggard; consider her ways, and be wise' (Proverbs 6.6, King James version). Deborah Gordon (2010), no sluggard she, of Stanford University, has been going to the red harvester ants and considering their ways in Arizona for 30 years. She is particularly interested in their capacity for self-organisation in the absence of central control. The queen is merely a reproductive machine – 'the ovaries of the whole colony' – and does not give orders. The worker ants have at least four different jobs. While doing a particular job, they produce a job-specific type of pheromone. These are volatile hydrocarbons that change the behaviour of other ants in the same nest.

Ants use these pheromones to communicate with one another. Wherever they go, they leave pheromone traces. When an ant encounters another ant, it samples its pheromone by touching it, and so receives a cue for the exhibition of a particular type of behaviour. The pheromones do not last long, as they quickly evaporate – indeed, this is one of their most important properties, because it clears the slate, as it were, for the next day's activities.

The ratio of the numbers of ants doing different things is maintained more or less constantly. Obviously, ants cannot count, but what they can do is secrete different types of pheromone, depending on the task they are doing. Thus, an ant that is foraging will spread forager pheromone on all the ants it meets. If an ant bumps into a large number of foragers within a certain time frame, and into very few cleaners, it will change its job to cleaning. Thus, the different sorts of ants within a colony can carry out a variety of tasks that maintain the colony in a high degree of order. 'Overall,' says Gordon, 'the red harvester ant demonstrates a remarkable ability in social cognition, cost-benefit analysis, and behavioral economics'.

The nest self-regulates. It does not matter if an ant with a single objective marches straight by a food source to which it has not been directed. It is an imperfect system, but then it doesn't need to be perfect, as resources are continually renewed by flood and wind.

Many organisms have crossed the line towards social behaviour. Bees and wasps belong in the same taxonomic group as ants. Bees in particular have a rich palette of pheromones that allow them to maintain a highly organised hive. They do better than ants in foraging because they fly and their behavioural transmission of information – the waggle dance – gives information about direction, distance and quality of a food source. Termites belong to a different group of insects altogether, but their organisation is also based on pheromones.

Many vertebrates, and especially humans, are social animals. Compared with the insects, they have greater degrees of intelligence and consciousness. Humans have high levels of both, which gives us flexibility in ways of doing things. In spite of this, there are so many, many similarities in the ways different groups of humans construct societies that this suggests that our biology is unconsciously dictating how we go about things.

The extended family, to take one example, is an institution that greatly facilitates the rearing of offspring whose continued immaturity is measured in years. Humans are far from unique in this. Cooperative breeding is common in birds – cooperative breeders include Australian magpies, white-winged choughs and blue wrens, where there is one breeding pair bringing up the young assisted by uncles, aunts and siblings. It is a characteristic of birds that live in a harsh environment (Stacey and Koenig 2006).

The human evolutionary journey from the origin of the first cell to the struggles of modern humans is part of the vast interconnected weave of Gaia. As for any species, it is story of boundaries crossed and collaborations established. In following this line of inquiry, the principal difficulties arise, ironically, from the social boundaries imposed by current ways of thinking in this, the second decade of the twenty-first century. One of these impediments is the tendency for dichotomous thinking – either this one *or* that one, not both – that has resulted from the 300 years of the scientific Enlightenment.

It is the Western reductionist tradition of separating the parts from the whole, the physical from the social, the rational from the creative and the individual from society that has created many of the boundaries encountered by humans on their journey. These are boundaries that have impeded the unearthing of the coherent and fully integrated story of life.

10

BECOMING HUMAN

In which we begin a fast-forward look at the human journey from tree-dwellers to moonwalkers and the potential expansion of Gaia into space.

Our evolutionary journey over the last 30 million years shows clearly our biological origin and that we are part of Gaia. Insignificant at first, we have made up for that in the last 500,000 years. Along our timeline we look for the characteristics that make us human: adaptability, tool use, intelligence – and empathy. These do not stand outside of biology but are part of it, supplemented by our own social evolutionary future.

It has already been remarked that being human is another way of being a fish. Not a flat, bottom-dwelling fish, but a goldfish, say. Everyone is familiar with a goldfish. It is an excellent swimmer, swimming with lateral flexures of a muscular body shaped like an aeroplane fuselage. It has pectoral and pelvic fins to provide steering and stability. Its gill slits allow a continuous respiratory flow over the gills as it moves forward, and its swim bladder gives it buoyancy so that it can 'hover' at different depths. This general design first appeared, in its basics, about 500 million years ago, and today, with a bit of pushing and pulling, it is a template that fits all animals with backbones (see Shubin 2008 for an excellent account of this process).

Goldfish and humans probably last shared an ancestor millions of years ago. Yet we still carry with us the signs that betray the fact that evolution, while often creating novelty, also tinkers with what has gone before. We are unmistakably fish-like. Our gill slits are gone, appearing only briefly during our development before becoming adjuncts to our immune system as tonsils. Instead, we breathe with an evolutionarily modified swim bladder, our lungs. Less efficient than gills, with their continuous flow, we have to breathe in and out through the same hole.

Our musculature and nervous system is segmented, based on the fish model. Our limbs are modified pectoral and pelvic fins. Our backbones, no longer horizontal, enable us to stand erect but are not well adapted to load bearing, as your first slipped disc will tell you. They do, however, free up our fins, so that notable fish, such as Leonardo da Vinci, could paint the Mona Lisa. It has been a long walk (swim?) to the Louvre!

The theme of this satirical poem, *Similar Cases* by Charlotte Gilman, illustrates the important truth of the opening paragraphs of this chapter. In evolution, a current state does not pre-empt a possible future state.

> Said the little Eohippus, 'I am going to be a horse!
> And on my middle finger-nails to run my earthly course!
> I'm going to have a flowing tail! I'm going to have a mane!
> I'm going to stand fourteen hands high on the psychozoic plain!'
> The Coryphodon was horrified, the Dinoceras was shocked;
> And they chased young Eohippus, but he skipped away and mocked.
> Then they laughed enormous laughter, and they groaned enormous groans.
> And they bade young Eohippus go view his father's bones.
> Said they, 'You always were as small and mean as now we see,
> And that's conclusive evidence that you're always going to be ...'

Eohippus, the 'dawn horse' was the small ancestor of the modern horse. Other 'similar cases' dealt with, not very flatteringly, in this poem are those of the 'anthropoidal' ape and Neolithic man. Like *Eohippus*, humans started off small. Our ancestors were squirrel-sized tree-dwellers. But we eventually became atom splitters.

It has been a 40-million-year journey extending across three geological epochs: Palaeocene (66–57 million years ago), Eocene (57–34 million years) and Oligocene (34–23 million years). Some have christened our present era the Anthropocene, marking the influence of our species on the biosphere. As the previous epochs lasted an average of 15 million years and *Homo sapiens* is at the most only about 400,000 years old, this seems to exhibit a marked degree of anthropocentrism.

Charlotte Gilman's poem paints an amusing picture of the little *Eohippus* boasting about his future as a modern racing thoroughbred. The 'heavy aristocracy' of the time were slow, lumbering mammalian forms: *Coryphodon* was a swamp-dwelling vegetarian about the size of a rhinoceros, while *Dinoceras*, with a remarkable, large, lumpy skull, was a distant relative of the elephants. *Dinoceras* now rejoices in the large, lumpy name of *Uintatherium*, because its original fossils were found in the Uinta Mountains in Utah.

The percentage of oxygen in the Earth's atmosphere has varied over time. It reached a record peak of around 30 per cent about 300 million years ago. In the subsequent 100 million years, the oxygen concentration drifted downwards until, about 200 million years ago, it bottomed out at 15 per cent and then drifted higher to today's value of 21 per cent. During the good times of high oxygen levels, animals with high oxygen demands appeared. These included many small mammals and a large group of small dinosaurs, ancestors of today's birds (Ward and Kirschvink 2015). At the beginning of the Eocene, a mammalian revolution was just beginning.

Most of the modern orders of mammals, or, at least, their oldest known fossils, appeared in the early Eocene. They include the Artiodactyls, 'even-toed' ungulates that walk on their third and fourth toes. This large group includes all the deer and deer-like animals. They are also related to the whales that took to the sea about this time. In contrast, the Perissodactyls, the 'odd-toed' ungulate group, that walk on their middle toes, include the horses and their relatives. To quote again from the poetry of Charlotte Gilman that started this chapter:

> Said the little Eohippus 'I am going to be a horse!
> And on my middle finger-nails to run my earthly course!'

At the beginning of the Eocene, many of the newer mammals were much smaller than those left over from Palaeocene. Modern groups appearing at this time include the bats, the elephants, the rodents and our own ancestors, the primates.

It would be nice to think that primates originated in a gloriously unique ancestor; sadly, this is not the case. Modern studies suggest that the descendants of the earliest primate-like animal was an insectivore that led, on the one hand, to tree shrews, rodents and lagomorphs (Rabbit and his friends and relations) and on the other … to us. Imagine something like a squirrel with its grasping hands and feet. These first primate-like mammals are found as scarce and fragmentary fossils that are about 60 million years old. They appear to be adapted, like tree shrews, to a life in the trees in warm, moist climates. Being tree climbers, they had the first two essentials of arboreal life – good stereoscopic vision and grasping hands and feet that were adapted for gripping branches and would, much later, grip tools.

The fossils of our closer relations, the first true primates, appear about 55 million years ago in North America, Europe and Asia. With larger bodies and bigger brains, it has been suggested that, as brains devour oxygen, increasing brain size was also a response to the increasing levels of oxygen in the atmosphere. If so, it is an example of an important evolutionary innovation that owes its origin to the activities of all green plants and ancient cells, whose photosynthesis continues to produce oxygen as a waste product.

The extinction event that occurred around the beginning of the Eocene was relatively small compared with the great extinctions of previous epochs. It affected mainly marine animals, including the ancestors of modern whales. A cooling climate did the damage, perhaps due to a lot of volcanic activity or to a succession of meteor strikes. Before the extinction event, an archaic one-toed and two-toed ungulate fauna held sway. There were also rodents and some very ancient primates still at the 'tree shrew' level of evolution. The Smithsonian's Erin Wayman (2012) gives an excellent general account of five early primates.

After the crash, it turned out that the ancestors of the rhinoceroses, pigs, hippopotamuses and modern ruminants had made it through the bad times, together with the ancestors of modern rodents, dormice and hedgehogs. So too did primates, in the shape of the family that gave rise to modern tarsiers and the 'dawn monkey' (Beard 2004), weighing less than 200 g, which was discovered in China in the 1990s. Towards the end of the Eocene, our distant cousins, the New World monkeys, emerged, together with our probable ancestors, the flat-faced Old World monkeys. The primate-like mammals were thus insignificant during the changes in the terrestrial ecology immediately following the extinction event.

The ancestral 'almost-monkeys' began to resemble more modern primates as the Eocene unfolded. Brains and eyes increased in size, with an emphasis on binocular vision, and faces became even flatter. The hole in the skull through which the spinal cord joined the brain moved forward, allowing the skull to be held in a more vertical position. This suggests that they were beginning to hold their bodies erect while hopping and sitting, like modern lemurs, thus freeing their hands to tend infants and hold tools.

By the end of the Eocene, about 34 million years ago, many of these early mammals had themselves become extinct. It took a further 10 million years before the first recognisably apelike human ancestor appeared. And about 20 million years after that, in 1909, part of a fossil jaw was found.

To digress slightly, two criteria must be satisfied to obtain an accurate indication of the distribution and diversity of fossils. You must have a good idea of where to look and there must be a lot of searchers. These criteria were far from being satisfied until after the middle of the twentieth century. The hunt for fossil hominids that was conducted before World War II was remarkable, however, for the insightful individuals who followed their hunches. Eugene Dubois, Raymond Dart, Robert Broom and various members of the Leakey family stand out. The fact that they were successful testifies both to the quality of their hunches – and their great persistence.

Dubois's great discovery came in 1891 – inevitably, he thought of it as a missing link, but not in those words. Instead he described it as 'a species intermediate between humans and apes' and accordingly named it *Pithecanthropus erectus:* the upright half ape/half man. It was not, however, human. At the time, *Pithecanthropus* was a celebrated find, for finds of any sort were in such short supply. Piltdown Man, the name given to the doctored bits of skull and jaw that were 'found' in a quarry, was a famous fraud perpetrated about this time.

The jaw that was found in 1909, by an unnamed gold prospector in Kenya, was better fitted to the Victorian image of the 'missing link'. Twenty-two years later, in 1931, at Lake Victoria in Africa, Arthur Hopwood and Louis Leakey found fossils belonging to three individuals whose jaws were similar to the 1909 discovery. The find was given the name *Proconsul* by Hopwood in 1933, after a series of performing circus chimpanzees all named Consul. At the time it was considered to be the oldest hominoid fossil ever found, but now the view is that it belongs to a sister group. There are several related species known from the fossil record and they all share the shortened snout and grasping hands and feet. *Proconsul* probably went on all fours, as it had a backbone unsuited to bipedalism. It was obviously successful, as its fossils have been dated as recently as 10 million years ago. It is a strong candidate to be ancestor of other apes, if not our own line.

A visit to Olduvai Gorge in the Great Rift Valley of Africa is as awe-inspiring as one to Lake Mungo in Australia, Stonehenge in England or Chartres Cathedral in France for anyone who has a sense of the past. It is an otherwise unremarkable ravine in dry scrubland, scoured out by a stream that has run through it for thousands of years. Consequently, its exposed geology has made it a particularly rich fossil ground. Olduvai was the hunting ground of Louis and Mary Leakey and family. Not far away is Laetoli where Raymond Dart and Robert Broom were making many interesting discoveries. A list of fossils illustrating human evolution may be found on the Internet so it is not necessary to go into great detail here, especially as the various species have been subject to much revision.

It is clear that, since the time of the writing of *On the Origin of Species*, we are now far from a scarcity of intermediate forms in our ancestry. In fact, we have an embarrassment of riches, and with the riches comes confusion that is only now being sorted out. The fossils tend to fall in to four groups, one of which includes ourselves (Figure 17). The groups are: the ardipithecines, the earliest, followed by the australopithecines and then three species (so far) of paranthropines. Our own group, the *Homo* group, comprises everybody whose remains have been dated to the last million years. An excellent account of human evolution is to be found in Ayala and Cela-Condi (2017).

All of the great apes – and us – are now to be found in the superfamily Hominoidea (Figure 17). The superfamily is split into two families, the Hylobatidae, which includes all the gibbons, and the Hominidae. Nothing more needs to be said here about the four genera of gibbons, today found only in Madagascar.

The family Hominidae (our family) has been split into two subfamilies, the Ponginae and the Homininae. The Ponginae contains only the Orang-Utans. We therefore number ourselves among the living Homininae, an honour we share with the gorillas and the chimpanzees. The Homininae have been divided into two tribes. In the larger one, the Hominini, we find ourselves and our nearest neighbours, the chimpanzees, while the Gorillini contains only (you guessed it) the gorillas.

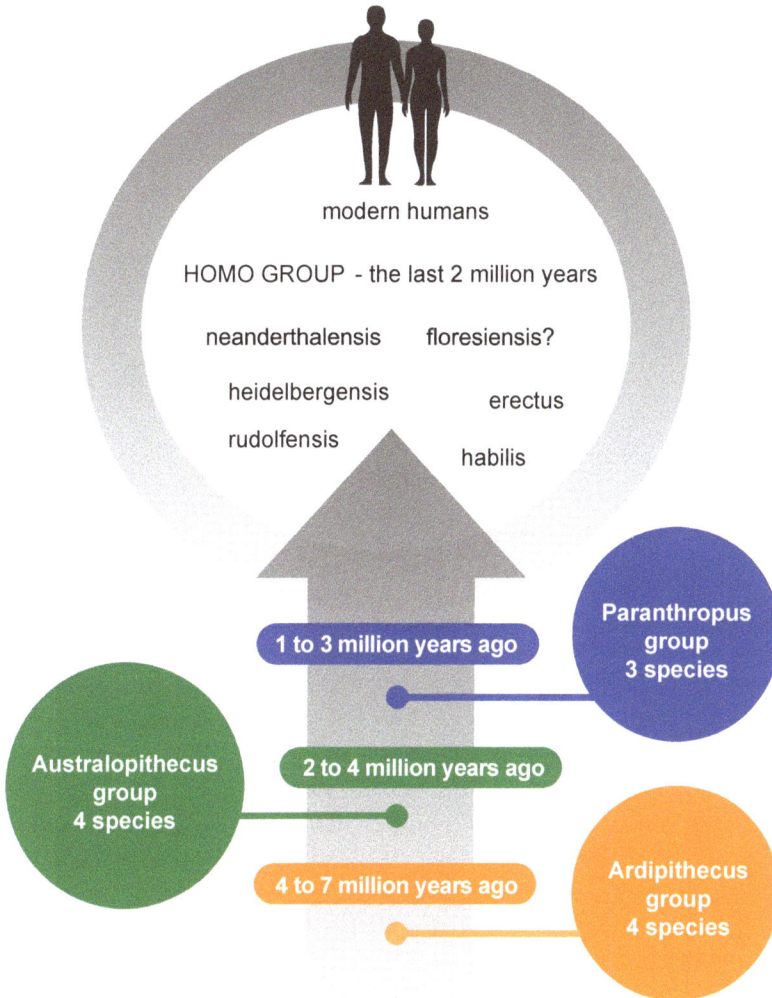

modern humans

HOMO GROUP - the last 2 million years

neanderthalensis   floresiensis?

heidelbergensis   erectus

rudolfensis

habilis

1 to 3 million years ago

Paranthropus
group
3 species

Australopithecus
group
4 species

2 to 4 million years ago

Ardipithecus
group
4 species

4 to 7 million years ago

Figure 17. Modern humans and our ancestors.

Finally, within the Hominini, there are two living genera, *Homo* and *Pan*. *Homo* has one extant species – us, *sapiens,* while *Pan* contains two species. *Pan troglodytes* is the famous chimp of zoo tea parties who originates in the Congo. *Pan paniscus*, the bonobo, is a smaller chimp from further south. Bonobos are our nearest evolutionary relatives, with whom we share almost 99 per cent of our DNA. They are therefore of particular interest to us.

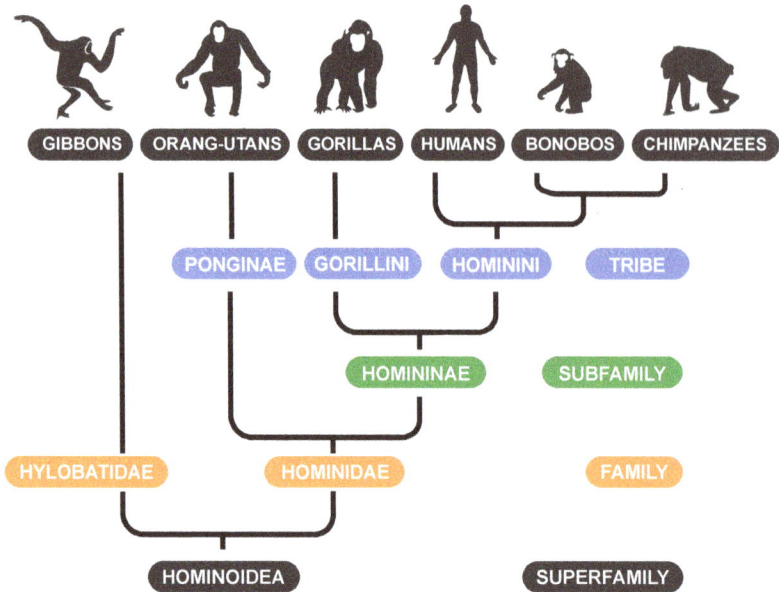

Figure 18. The Hominoidae superfamily.

Returning to the group to which we belong, now known as the Hominini, among the earliest members are *Homo rudolfensis*, known from three individuals, two of which were found in Turkana, in Kenya; and *Homo habilis*. Their fossils are about 2 million years old. *H. rudolfensis* had a brain volume above the upper end of the range for known *H. habilis* fossils. At the moment, these two are designated separate species but as 'one swallow does not a summer make', so one fossil, or even two, does not necessarily make a species. There remains a question mark over *Homo rudolfensis*.

To save an irritating proliferation of italics, in what follows, hominins will be referred to by their species or 'surnames' – as we might refer to Darwin or Huxley. Habilis was the 'handy man' and its (to avoid saying his or her!) fossils are known from east Africa. It was apparently brainier that the earlier hominins but was apelike in its prominent jaw and relatively long arms. It was named *habilis* as it was thought, at the time of its discovery, that it was the earliest maker of stone tools. It was a small hominin, not quite 1.5 metres tall and weighing about 30 kilograms. Since then, even earlier tools have been discovered, that have been made by … who knows?

About 1.5 million years ago, Erectus appears in the fossil record. Some authorities think that Erectus is not one but four species. In any event, its type was widely distributed in Africa, Georgia, China and Indonesia, and it was evidently a traveller – not as we might travel, to a specific destination or 'to boldly go where no hominin has gone before' – but little by little, generation by generation, as it extended its range. It was roughly the same size and proportions as modern humans. Significantly, there is fossil evidence that Erectus cared for its weak and its dead.

Erectus was a denizen of the savannahs, a way of life more conducive to movement on the ground, perhaps even to nomadism, than life in the trees. The larger body and longer legs made journeying for longer distances possible while the larger brain was equipped with an intelligence able to cope better with the vicissitudes of different environments and to make better tools. It has also been suggested that the Erectus diet included meat, which provided enough energy for the bigger brain and also permitted survival in temperate latitudes in winter, when food plants were dormant.

From the dates, almost everyone seems to agree that Erectus was gone about 400,000 years ago. It may be that its relict persisted for another 350,000 as the 'hobbit', Floresiensis, but the jury is still out on that one. Perhaps Floresiensis, imprisoned in the evolutionary hotbed of Flores Island, is a dwarf sister species of Habilis.

Anyway, everyone agrees that Floresiensis was a sideshow, but this is not the case for Heidelbergensis nor Neanderthalensis. These two hominins appear in the fossil record in the last million years. Heidelbergensis was the first to arrive, about 700,000 years ago while Neanderthalensis exists in the fossil record from 400,000 to 40,000 years ago and is with us today – in our DNA.

About 2 million years ago, Erectus, still in Africa, gave rise to the line that led to Heidelbergensis. This ancient Erectus stock continued in Africa, but around 700,000 years ago Heidelbergensis began moving to the Middle East and then westwards into Europe. They, or Neanderthals, were in France and Spain about 100,000 years ago. A group of humans, from Denisova in Siberia, are of great interest. They are not yet granted species status but named from a few remains found in a cave. One fossil find, a bone from a 13-year-old girl, is remarkable because analysis of DNA obtained from it shows her to be a hybrid produced by a mating between a Denisovan and a Neanderthal.

DNA analysis shows that modern humans, Neanderthals and Denisovans, had an ancestor in common about 600,000 years ago. Further evidence from mitochondrial DNA analysis suggests that the Denisovans were the result of an earlier migration out of Africa, distinct from the later out-of-Africa migrations associated with modern humans and Neanderthals, but also distinct from the earlier African exodus of Erectus.

In order to simplify discussion, it has been customary to divide our species into 'Archaic' and 'Anatomically Modern' humans; those forms are used here. Archaic humans have a greater physical resemblance to their heavier ancestors, and some argue that the latter represent subspecies of Sapiens – thus, *H. sapiens neanderthalensis* and *H. sapiens heidelbergensis* – or even separate species. Anatomically modern humans have lighter builds than their archaic ancestors and large brains that vary in size between populations and sexes but average 1300 mL.

Sapiens is first recorded in the fossil record in Morocco about 300,000 years ago. Remains have been found in Israel that are 100,000 years old, and slightly younger fossils have been discovered elsewhere in Eurasia. In contrast, the famous Cro-Magnon discovery provides earliest evidence of anatomically modern humans in Europe. The bones and the associated painted caves of Lascaux are only 30,000 years old. (For comparison, the skeletal remains from Lake Mungo in Australia are 40,000 years old.) Cro-Magnons were evidently strongly built. The high forehead had only slight brow ridges. The face was short and wide with a prominent jaw. The brain capacity was about 1600 mL, larger than most modern humans.

At some time during these journeys, they passed through a metaphysical barrier that Midgley (2004) marks as the transition from hominin to fully human: they learned to tell stories. Stories are really imagined futures. So are lies.

The whole of biology is beset with 'lies', with plants and animals pretending they are not there, or that they are something else, that they are dangerous, that they are poisonous, that they are fitter (in an evolutionary sense). In humans, children as young as two years old are capable of spontaneous lying and such behaviour increases dramatically by the time they are three. Children with better cognitive ability are capable of telling better lies (Evans and Lee 2014). As children develop, their ability to balance more than one 'reality' in their heads becomes easier.

Modern humans can thus visualise numerous possibilities, options that offer a large range of outcomes that vary from 'very good' to 'disastrous'. If you were to plot them as a normal distribution curve, you would get the 'disastrous' ones at one end and the 'brilliant' ones at the other. The hump in the middle would be occupied by the 'not too bad and not too good' – average, in fact. Setting aside those, like the winners of the Darwin Award, whose final disastrous choices remove them from the human gene pool, and Nobel Laureates at the other end, most people select from a range of 'imagined stories' whose outcomes are somewhere in the average hump. There are far more 'not so good' options than really good ones, and far more 'not so good' choosers than 'good' choosers. 'Not so good' outcomes, therefore, are more likely to be chosen, and to be tested either evolutionarily or socially. The sheer complexity of modern society created by billions of people has also created an enormous range of future options of the 'not so good but not disastrous' category.

The ability to tell good stories clearly has considerable survival value within human social systems. Sociologists Barthes and Duisit (1975) have argued that, despite the diversity in how it is told, there is only one human story: a villain and a hero searching for a treasure. This can even be applied to the evolutionary story. In an anthropologist's overview of evolutionary advances, say, a species is a hero if it brings the evolutionary journey closer to humans; a villain if it moves away from a hopeful extension (for example, the Neanderthals). The treasure is, of course, becoming human.

Ability to make choices about the future follows the same rules as other biological phenomena and conforms to a normal distribution curve. Those in the population who are able to make 'good' decisions will occupy one small tail, while the 'not so good' deciders will occupy the huge hump clustering around the mean. The other tail is occupied by the 'disastrous' choosers. Given that the population of Earth is 7 billion, the hump will comprise huge numbers of people, most of whom will be intelligent, and many of whom will be influential and may have an interest in promulgating 'not so good' scenarios.

A good leader, who understands their group's circumstances, will be able to extrapolate from them, see ways in which they might be improved and then impart that vision to others. Archaeological evidence suggests that hominins other than Sapiens had this ability.

Rendu et al. (2014) describe a Neanderthal burial at Chapelle-aux-Saint. It appears to have been a naturally formed trench that was modified in various ways to make it a suitable receptacle for the body. The authors concluded that these constitute convincing criteria for establishing purposeful burial. Studies of 28 human skeletons in Atapuerca in Spain suggest that Heidelbergs might have been the first hominins to bury their dead (Carbonell et al. 2006). Burying one's dead and using paints implies envisaging futures that are different from the real one or perhaps indicates respect and affection for the dead person. It may also reflect the recognition of a cause and effect relationship, with hygiene the main consideration – burial helps to prevent disease from a decaying corpse. The 'Midgley Point', the moment at which these practices arose in hominins, must predate this. Steven Mithen (2005) believes that both Neanderthals and Heidelbergs had acquired a pre-language system of communication. Red ochre, commonly used in painting and decoration, has been found at the Terra Amata excavations, recently dated at 230,000 years ago in the south of France, although no forms of art nor sophisticated artefacts other than stone tools have been uncovered.

Ergaster, an African form of Erectus, is thought to be the first hominid to vocalise, although the evidence on which this is based involves much supposition about airways and, to an inexpert reader, conveys little more than the possibility of modulated grunts. A better bet may be Heidelbergs and Neanderthals, who developed more sophisticated culture. The morphology of the outer and middle ears of these hominins suggests they had an auditory sensitivity similar to modern humans and very different from chimpanzees. According to Mithen, they were probably able to differentiate between many different sounds. Others have compared the structure of the vocal apparatus of five Heidelberg fossil individuals with modern humans and chimpanzees. Recently, a bonobo has learned to communicate, using geometric symbols representing words, without the arduous training required by the famous 'talking apes' of earlier studies.

These observations together suggest an origin for human speech about half a million years ago. Presumably, the ability to speak is a prerequisite for the telling of stories, so our capacity for mendacity has a considerable ancestry. The interesting question is whether it arrived before or after the development of an ethical sense to make it respectable.

On looking back, it seems inevitable that something like genus *Homo* would emerge. We see all about us trends that appear to point the way to bigger brains. Looking forward from 3 billion years ago, however, it is by no means obvious that the emergence of genus *Homo* was inevitable. Given our history, perhaps there is room for only one major intelligence per planet. Perhaps the qualities of intelligence and consciousness are such that an evolving planet only gets one shot at it at a time. There are some good understudies in other groups in our evolutionary bush waiting in the wings, although it might take them a few million years or more after we depart the stage.

> There was once a Neolithic Man, an enterprising wight,
> Who made his chopping implements unusually bright.
> Unusually clever he, unusually brave,
> And he drew delightful Mammoths on the borders of his cave.
> To his Neolithic neighbours, who were startled and surprised,
> Said he, 'My friends, in course of time, we shall be civilized!

This is the last we shall quote from the witty Charlotte Gilman, for at the end the poem turns dark and pessimistic!

It is time now to take a look at our own little taxonomic cluster, the Hominini. There are three of us left in it, the three chimpanzees, as our other *Homo* cousins have long departed.

The point in time at which *Homo* separated from the two chimpanzees is clouded because the extent of hybrid speciation is not known. Hybrid speciation occurs when two similar but not identical species successfully interbreed. It has not been thought to be a common occurrence, and conventional wisdom suggests that the offspring are usually sterile – think of mules and ligers. It has, however, obviously occurred in the evolution of modern humans since there are Neanderthal and, recent evidence suggests, Denisovan genes in the modern human genome.

11

INHERITING THE EARTH

Having established a zoological ancestry for humanity, countering the fundamental religious argument that we are unique and outside of nature, we follow the paths that humans have travelled out of Africa.

Originating in Africa, humanity undertook a series of migrations, spreading through Europe and Asia, finally arriving in Australia and the Americas. At some point a suite of evolutionary changes occurred that separated one line of hominins from their close relatives, with unforeseeable consequences.

Perhaps the need to accommodate to the rapid – from an evolutionary point of view – enlargement of their brains or to some other change in their environment, such as diet, our ancestors began to enter the world in an immature and defenceless condition. At another time and in another place, this could have been an evolutionary calamity, but it was accompanied by other changes to protect the newborn, including intense maternal care. A process of neoteny, not uncommon in evolution, had begun.

The change proved not to be the disaster it might have been. The babies had the potential to grow large brains, unhampered by premature hardening of the brain case. Early birthing also allowed a baby's head to fit better in the birth canal. The downside was that the hominins had to spend more time and effort in rearing their young successfully. We are now remarkable among animals for the years of work we put into our children.

It turned out in the end to be a great evolutionary success. The changes led eventually to all we take for granted in our human condition. Of course, one thing does not change in isolation. The need for prolonged child care created a selection pressure for greater cooperation within the family group to provide for and protect nursing mothers. And there were physical changes as well, developmental changes that once happened in utero that now happened in early childhood. These characteristics, once isolated from the environment, were exposed to it and subject to its influences.

Our ancestors, it seems, were the first hominins to show this form of evolution, called neoteny. It is essentially the retention of juvenile features into adulthood. It is a phenomenon that exposes persistent embryonic features to direct natural selection. Other primate species do not experience this as intrauterine development produces newborns with more advanced muscular development.

The idea that *Homo sapiens* is a neotenic chimpanzee is an idea of some antiquity. In 1929, a Dutch anatomist, Lodewijk Bolk, first drew up a list of the probable neotenic characteristics that we, as adults, display:

1. A rounded, bowl-shaped cranium to house a brain that will grow to about four times its volume at birth.

2. A juvenile face that lacks the brow ridges and the heavy jaw development that is characteristic of adult apes.

3. The spinal cord enters the skull at a point vertically underneath it and does not move towards the back during development, thus making an upright stance possible.

4. Late closure of skull sutures and persistence of cartilaginous bone ends (epiphyses) into early adulthood.

5. In women, the vaginal canal does not rotate backwards, as it does in chimpanzees, but remains directed forward.

6. A big toe that remains aligned with the other toes and does not rotate sideways and become opposable, like a thumb.

Compared with our simian cousins, our skeleton is under-developed at birth, with cranial sutures not yet joined, and uncalcified ends to the long bones which do not fuse together until puberty and a 'straight', rather than a grasping, big toe. An unfortunate legacy, a backbone better suited to walking on all fours than standing upright is still a human drawback today.

In women, the reproductive tract does not rotate rearwards, as in apes. Its more forward position facilitates face-to-face sexual and other communication. All this was accompanied by a period of intense social evolution that paved the way for collaborating communities. It must be said that this appealing concept is not without its critics. Shea (1989) considered the evidence, and concluded that there was a case, but it was 'relatively weak' (which might be interpreted as 'quite strong' depending on your point of view!).

So now, equipped with a bipedal posture made possible by longer legs and changes to the skull, combined with a Palaeolithic tool kit that included stone points, hand-held axes and fire-hardened spears, ancient humans were ready to leave Africa. Recent fossil evidence from Morocco suggests that archaic *Homo sapiens* were in north Africa as much as 300,000 years ago (Hublin et al. 2017). According to Groves (pers. comm.) 'modern' humans appeared first in Africa and began their migrations into Europe and Asia about 80,000 years BP (before present). Evidence for this comes from 'modern' human fossils found in the Middle East, suggesting the trip was made from the Horn of Africa to the Arabian Peninsula by rafts or by island-hopping, or even walking, when the sea level was about 70 metres lower than it is today.

This was a trip that other non-*sapiens* hominins had made much earlier. For example, the diminutive 'hobbit' had already drifted as far as Indonesia. Argue and colleagues (2017) conclude that the most likely origin for the 'hobbit' is from an early *Homo* lineage, such as *Homo erectus*, who wandered all over Africa, but also made it out of Africa and into Asia. Its most recent fossils in Asia are dated at 140,000 BP. The Neanderthals were in Europe and Asia 400,000 BP, and persisted until 40,000 BP, making them partly contemporaneous with modern humans. Cohabitation probably accounts for the Neanderthal genes we possess.

The fact that the date for the exodus from Africa of *Homo sapiens* is set around 80,000 years ago is consistent with the new estimates for human arrival in northern Australia. This is ample time for humans to have made the trip given the built-in error in the age estimates. The very recent discovery of polished stone axes dated to 65,000 BP, and other artefacts, such as the presence of red ochre for decoration and/or ritual purposes, suggests that modern humans – Aboriginals – were in Australia at least 10,000 years earlier than previously thought.

The first scanty suggestions of human abstract thought come from well before the last ice age. The carefully laid-out body of Mungo man in Australia, from 40,000 years ago, supports this. More recently, even clearer evidence of abstract thought is found in the cave paintings in Australia, France and Spain. The images are of animals and food sources, and other symbols of social custom. It has been suggested that these cave drawings were intended to increase the luck of the hunters, to show respect for others or to invoke help from spiritual beings. A stone figure of a plump woman, the 'Venus of Willendorf', carved 28,000 years ago, might have served all three purposes. The careful burial of a man, at about the same time, decorated with ochre strongly suggests that those who grieved had the concept of an afterlife.

Exciting recent finds in northern Australia carry implications for our understanding of the spread of humans around the globe, but also pose some more immediate questions (Clarkson et al. 2017). If humans have been in Australia for 65,000 years, why is there no evidence for their spread to the Lake Mungo region over the following 15,000 or so years? Perhaps the Lake Mungo people just took a long way round or the time gap is simply due to too few archaeologists searching too large an area.

If you go to Lake Mungo – now dry for several thousand years – and stand on the old lakebed, you can gaze up at the huge sand dunes on its eastern shore. You are now looking at the so-called 'Walls of China'. The dunes are called lunettes and they are on the move. The prevailing wind is picking them up, sand grain by sand grain, and shifting them further east. As the top of the dune drifts away it uncovers a few old and twisted long-dead trees, stone tools and old fireplaces, all that remains of an ancient Aboriginal community. As you stand there, you will experience a silence broken only by the soughing of the wind and the hissing of the sand. For a European it is a numinous experience, akin to standing in the nave of a great cathedral. A conservative estimate suggests that the site may have been occupied for 50,000 years.

There are two important discoveries of early humans at Mungo. Lake Mungo 1 includes the earliest evidence, 40,000 years old, of a cremation. The bones are those of a woman, who has now been returned to her people. A second skeleton, Lake Mungo 3, an unusually tall and slim man, had been buried on his back with his hands folded over his abdomen. Controversy surrounded his geological age, but it is now accepted that he was more or less contemporaneous with the cremated woman.

Good stone for the manufacture of tools is scarce in the area, and most of that worked by the Lake Mungo people has been imported from elsewhere. Points, knives, hand axes and grindstones have been recovered. The grindstones served the same purpose as the mortars and pestles used in kitchens of today: they are large and flat, with a circular, shallow bowl worn into one surface by the continuous pounding and grinding of seeds and other plant material with a rounded stone reserved for the purpose.

If you now journey from Lake Mungo, and travel just over 900 kilometres north-east as the crow flies, and 20,000 years forward from the time of the Lake Mungo people, you will find yourself in Carinda, New South Wales. Carinda, a small town of perhaps 200 people, is close to the important archaeological site of Cuddie Springs.

Cuddie Springs is the site of yet another of Australia's dry lakes, a basin about 3 kilometres wide. Fossils of all kinds of animals have accumulated there, animals that, in life, were attracted to the shrinking waters and then died there from various causes. The fossil-bearing strata are 3 metres thick. The bottom levels are dated at 50,000 years ago. Signs of human activity, as well as the bones of large animals (examples of some of the famed Australian megafauna), are found at about the 30,000-year horizon. This fact alone should persuade people that the Aboriginals had cohabited with megafauna without immediately bringing about their extinction.

There is, however, little doubt they caught these large marsupials – or took advantage of them being mired in the lake margins – and ate them. A worn stone core was found, still showing traces of fur and blood, suggesting it was used in butchering. Preliminary results of the analysis of DNA extracted from the megafaunal bones indicate that they were those of a giant kangaroo (*Macropus titan*) and an even larger *Diprotodon,* some species of which were rhinoceros sized. Even more interesting, a number of grindstones were found, with traces of plant tissue and starch granules on their surfaces. Nearby, charcoal suggests the presence of cooking fires. It seems likely that these people were among the first bakers in the world.

Now fast-forward to the year AD 1800 and the arrival of the European explorers. In his fascinating and award-winning book, Bruce Pascoe (2018) has collected and annotated eyewitness accounts of Aboriginal society by European explorers. For example, Major Thomas Mitchell, in Western Australia, wrote in 1839:

> the grass is pulled and piled in hayricks, so that the aspect of the desert was softened into the agreeable semblance of a hayfield … we found the ricks or haycocks …

> … dry heaps of this grass that has been pulled expressly for the purpose of gathering seed lay along our path for many miles …

> … the seed is made by the natives into a kind of paste or bread.

This surely goes beyond simple gathering. It is agrarian industry, its origins clearly of great antiquity. Pascoe goes on to provide evidence for the cultivation of yams and other food plants, for irrigation, of ancient fish-traps, of stone structures and of the benefits of the so-called 'firestick farming'. In one poignant story he tells of a proud settler demonstrating European methods by ploughing up and down a slope. The local people, however, were not having their precious soils, soils that they had been conserving for millennia, washed away by runoff and as soon as possible reorganised the area so the furrows ran parallel across the slope.

One settler disparaged these activities as 'the accidental semblance of gardening' and we may imagine that this particular view was conveniently extended to deny other agricultural activities. It is hard to see at this distance what would have convinced the settlers that the locals were acting with purpose gained from long experience.

Eva Jablonka and Marion Lamb (2014) describe 'evolution in four dimensions' – genetic (based on gene variation), epigenetic (based on environmental effects on development), behavioural (based on cultural change) and symbolic (based on communication and learning). Lake Mungo provides evidence for all four having been in place for human beings in Australia for at least 50,000 years.

The evidence collected throughout this time supports the thesis that cooperation is the mainspring of evolution. Genetic selection has led to human groups with similar physical characteristics creating communities. Belief systems, informal and formal, provided the glue that held the early communities together. A belief held in common, however badly based on reality, ensured that the community learned to act in concert against a perceived existential threat, whether it was imaginary, as in the case of a wrathful god, or real, in the form of an angry cave bear. The capacity for belief seems to be innate in humankind; it is as pointless to tell someone to stop believing, whether in transubstantiation or little green men from Mars, as to ask somebody to tell their pancreas to stop producing insulin.

Epigenetics, the idea that inheritable changes can be brought about by the environmental context, is a more recently respectable study. Farming to ensure a continuous supply of food is one such example. The discovery of a form of farming by Aboriginals may predate by millennia the similar discovery in the Middle East. In so doing, they modified the selection pressures that were originally present in the environment and provided new opportunities for epigenetic evolution. The interaction between humans and their environments is an effective form of niche improvement.

This, then, is the period for humans when the nature of evolutionary change entered a new dimension. Up to this point, the inheritable changes, on which Darwinian evolution depends, have depended on the modification of pre-existing genetic programs. With the development of social and cultural environments, *Homo sapiens sapiens* created a very different evolutionary milieu. It was one that was facilitated by language and all the forms of collaboration provided by a mutually cooperative society: as individuals living and working together, as communities with deep attachments to their land, and in structured organisations.

Everyone accepts that animals have 'behaviour'. At what point does complex behaviour become 'culture', with its sophisticated forms of expression, symbols and ritual? In the northern hemisphere, farming, a cultural phenomenon, is assumed to have started about 12,000 years ago in Mesopotamia, well after the recent Australian discoveries. However, culture does not arise fully-fledged. There must have been 'proto-farmers' attempting to improve their land, their evolutionary niche, by trial and error. Even in medieval times in Europe, farming theory was so rudimentary that often the most valuable thing a farmer could leave his son was a well-stocked dungheap. Without this important resource to replenish depleted farmland, the family was in danger of starving.

We are comfortable with the idea of animal behaviour, but can animals have culture? Animals can learn. Blue tits in Britain are famous for learning to open milk bottles to get at the cream (Aplin et al. 2013). It presumably started out with the birds' random explorations of a new environmental factor – the closed milk bottle – in their normal environment. In whatever way it came about, the lid was breached, the bird got the cream and this happy outcome reinforced the behaviour. Because the blue tit lives in family groups, the new skill, of opening bottles, was quickly learned by others in the group. Juvenile females proved more likely to learn the trick. Had this habit been allowed to persist, it is likely that evolutionary changes would have occurred, perhaps to the birds' digestive physiology. Without the actual presence of the appropriate bottles, the birds had no means of passing it along to future generations. Sadly for the tits, the bottles were replaced by others with better seals. The new habit died out. It had been behavioural, tied to a very specific sort of bottle top, rather than culturally applicable to all bottle tops.

Consider now a group of Japanese macaque monkeys of Koshima Islet that has long been studied by behavioural scientists (Visalberghi and Fragaszy 1990). The monkeys live close to a warm spring. In winter these animals developed the habit of going swimming to keep warm, especially the mothers with their babies who thoroughly enjoyed frolicking in the water. One of these mothers was particularly clever. If fruit or vegetables were left on the sand for her, she would pick them up and take them to the water's edge and wash the sand off before eating them. Other female monkeys observed her doing it, and they all started washing the sand off their food before they ate. The young monkeys learned to do likewise, and this trick was added to the standard behaviour of the animals. It is interesting that the males did not learn to carry out either process, perhaps because they were less social than the females.

It is likely that there are many other monkey tricks that are potential precursors of human behaviour. Non-human primate archaeology is a new field that is providing insights into monkey behaviours that have respectable histories. Many monkeys use stones to crack nuts and they have favourite stones that they use as hammers and anvils. In so doing, they occasionally produce sharp-edged flakes that resemble the flake tools of early humans, although there is as yet no evidence that the monkeys use them in the way that early hominids did.

The next story is an example of learning as part of the acculturation of a human being. It is 1948. A 12-year-old boy is looking forward to an afternoon of 'stinks' in the chemistry laboratory at school. The schoolmaster in charge has promised that the experiment will be 'to burn metals in air'. The boy considers that this is both worth learning and fun. The society in which he lives confirms that it is worth learning because it is in the course notes. He takes his place at his workbench and watches as the chemistry teacher carries out the experiment, explaining as he goes along. Now, says the teacher, you do it. The teacher wanders around, giving advice and answering questions and under his watchful eye the boy carries out the experiment successfully.

The boy has learned a number of new skills in carrying out the experiment, skills that will come in handy in next week's lab session and, maybe, for a lifetime after. He writes the experiment up in a laboratory notebook to which he can refer if necessary and which, ideally, anyone else could use to repeat this experiment.

This process differs from the experience of the brainy macaque in a number of ways. First, inherited social conditions ensured that the laboratory, the teacher and the boy were all there at the same time. The boy and the schoolmaster could speak to each other, not only in the same language, but also using highly specialised terms of the science. The schoolboy had enjoyed a glimpse of a subculture of the main culture to which he belonged, the discipline of chemistry. In the language of his own boyish subculture, however, chemistry was 'stinks' and things excellent were 'trick super!' The laboratory itself spoke of a rapidly fading past subculture, with bottles labelled 'caustic potash', 'vitriol', 'blue vitriol' and 'hydrargyrum' (potassium hydroxide, sulphuric acid, copper sulphate and mercury).

The master had his own subculture, with the language of, fortunately, modern chemistry. He was there, not just to produce a single change of behaviour in the boy, but to play his part in educating a community. He was also a model for a technical performance and a prestigious member of their shared society. The whole process was designed to hand on the socialisation as well as the expertise. When the boy went home that night, he used the language of the main culture to explain to his parents what he had done at school and showed them the written symbols in his lab book. In a couple of decades or so he could have shown it to his own 12-year-old son, who would have understood it. This is cultural learning.

The transfer and acquisition of cultural learning provides an environment for inherited change totally different from that offered by the natural world. Darwin did acknowledge this in his treatise on the emotions although he didn't take it any further. Changes in thinking, use of language, standards of behaviour and values cannot be equated to genetic inheritance, although they can lead to evolutionary changes by contributing to the wellbeing and survival of a greater number of people than otherwise might have done so – an example of niche improvement.

The boy's story was drawn from the cultural context of one of the authors of this book in mid-twentieth-century lower-to-middle-class England. There are many different cultures, all offering different evolutionary possibilities. Sadly, there is also the possibility of toxic cultures – the Stalinist purges, Nazism, the Pol Pot regime – but these fortunately have shown a tendency in the past not to survive for very long. There is a phrase, 'cultural memes', for words such as gene, sustainability, and learning that Dawkins and other writers argue are the cultural equivalent of genes (Heylighen 1998). Some memes, like some genes, have negative consequences.

The clever macaque may pass on her idea to other monkeys – monkey see, monkey do – especially her young ones. However, we must presume she cannot generalise and use her discovery in other contexts. Nor can she tease out the original principle from her empirical understanding of washing sand off her fruit – the idea of specific gravity, and that sand has a higher specific gravity than most fruit and will sink in water – because that requires the symbolic representation supplied by language. She may be no better off at thinking than the blue tits. Changes in human thinking have been variously described: by positivists as recombination of memes, by geneticists as evolutionary advances, by anthropologists as the 'golden bough', by the classicists as the great chain of being and by science fiction authors in as many forms as imagination can take.

Organic evolution was, in Darwin's time, a new way of thinking, and although he shattered the boundaries of Victorian thought, he was confined both by what he did know and what he could not know. We too are similarly constrained because we can only discuss evolution in terms of the ways of thinking of our own times.

The same is true for studies of evolution of the mind. Physical differences mark Neanderthal, Cro-Magnon and modern humans. Early palaeontologists could not resist the idea that the three represented an evolutionary sequence. This idea has now been discarded and they are understood to represent parallel evolutionary lines with a common ancestor. Even the supposition that the Neanderthals were less intelligent than modern humans, based on skull shape, has been dismissed. It is now generally accepted that *Homo sapiens* originated in Africa and different groups emerged from Africa at various times and spread across the world. As they travelled, each developed physical differences, in response to the environments in which they settled, that are still apparent today. In the modern era, movements of huge populations of travellers and asylum seekers are leading to intermarriage and a mingling of genetic information across a wider spectrum than ever before.

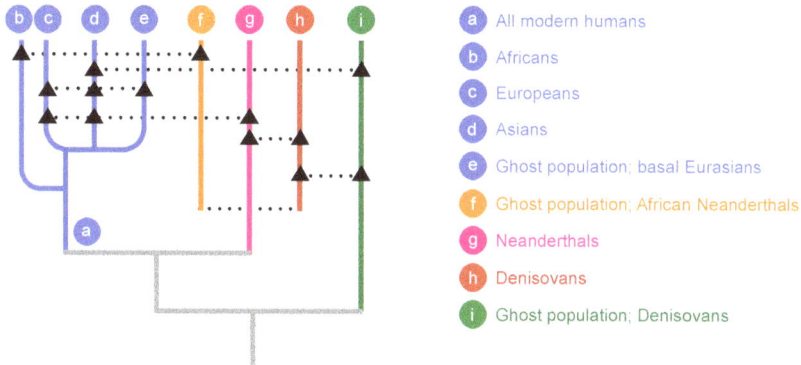

Figure 19. Ghosts in the DNA.

In spite of the differences, however, we are all one species if the standard definition – a group of individuals that actually or potentially interbreed in nature and produce fertile offspring – is accepted. Recent genetic analyses from fossil humans and from living humans belonging to defined groups, like Africans, Asians and Europeans, have suggested a human family 'tree' strongly reminiscent of an irregular network. A bone from the Denisova cave in Russia has yielded 'fossil' DNA. This young person is a perfect hybrid of Neanderthal and Denisovan; in other words they represent the immediate offspring of a mating between the two types of parents. Even more astonishing are the 'ghosts' in the DNA of various types of human.

A recent study, reported in *New Scientist* (Brahic 2018), describes the discovery of 'ghost' DNA from different sorts of humans (Figure 19). The 'ghosts' are of people who once formed coherent groups as distinctive as Neanderthals and Denisovans, but who are only known by their specific DNA and are unknown in the fossil record. In the figure, these 'ghosts' are indicated by dotted horizontal lines. Thus, modern humans (b) share DNA with a 'ghost' population of African Neanderthals (f), and so on. They are indeed true 'missing links' known only from samples of DNA. It says something about the inclusiveness of our ancestors. When two groups met, and each looked human to the other, they mated. Perhaps, like bonobos, copulation was used as a way of defusing potentially aggressive situations.

Language, spoken or written, is a principal driver of social evolution. Before there was written language there was an oral tradition. Modes of artistic representation had the same effect. The Pleistocene drawings of cattle in the Lascaux caves, the religious icons of the Renaissance, the realistic paintings at the turn of the nineteenth century, impressionists at the turn of twentieth and the abstractions of this twenty-first century both record and shape the societies of their times. The cultural effects of change can reside in the very identity of a place. Change it, or move the people, and a culture and its individuals change dramatically. The same degree of change was observed with the advent of the electronic era, when, according to McLuhan (1967), the medium became the message.

Yet another avenue of change is in the educational transfer from one generation to another, as in our example of the schoolboy. Here there is a formal connection between evolutionary, environmental and social change. In the twentieth century, the 'jug and glass' concept of education was common, where knowledge is poured from a reservoir (a jug or a schoolteacher) into an empty glass in the form of a small boy or girl, until they are full and able to empty it out again in written form in an examination hall. Now that has given way to 'discovery learning', and the view of the world has changed. Research has become an investigation into physical environments and the chemical structure of the units of inheritance. This in turn allows human interventions that change the structure of the environment and so influence the evolutionary direction of the era.

The cultural setting shaped by social learning is always a complex construction of law, education, beliefs, resources, defences, health and landscape. The problems posed by any historical period can only be completely understood by the people living in it at the time. Oral history, poetry, theatre and creative literature have been brought in to help this process in the past, as now we depend on the cinema, television, the press and social media. We can, however, only build our ideas of the evolution of the mind through archaeological discoveries and historical analyses. It is often a very chancy business.

There is much evidence that the evolution of *Homo sapiens* has not stopped. At a recent conference (reported by Pennisi 2016), considerable variation was reported in human DNA of the last hundred years or so. For example, a 'smoker's gene' that makes some people more vulnerable than others to the effects of smoking has become less common in the population, presumably because of the increased early mortality of heavy smokers. At the same conference it was reported that in traits controlled not by a single gene but by hundreds of genes, there were many small changes occurring. Among those are height, head circumference in infants and hip size in women. Many other changes were noted across the whole human genome.

Here are more examples of changes that affect human interrelationships that may bring about evolutionary change. Please keep in mind the statisticians' mantra that correlation is not causation! Similar variations with time in two separate sets of relationships do not necessarily mean that one is causing changes in another. One classic statistical joke is that, in our society, there is a positive correlation between the incomes of bishops and prostitutes. We must suppose that it is not that one has a direct influence on the other, but that both live in the same economic climate.

Late onset menstruation has been found to correlate with longer life. In Western society, however, menstruation is coming earlier and life expectancy is increasing rapidly. The incidence of pale skin, blond hair and blue eyes is correlated with a diet change. Perhaps the shift from fish and nuts to grain and meat with the onset of farming led to vitamin D deficiency and favoured those with paler skins. Mate selection in Africa is heavily influenced by the occurrence of steatopygia in women, the possession of large buttocks that evidently act as a fat store. This has survival value in a country that suffers regular famine. At the other end of the scale, the Western ideal of female body image reaches to the point of anorexia, even within members of the British royal family.

At the start of the fourteenth century, it is estimated that the world population was 450 million. By the end of the century it was down to perhaps 350 million, mainly due to the depredations of the plague. This is a stark reminder to us of the perils of being a species monoculture. People or potatoes, it is the same. If a lot of them grow together, they become vulnerable to disease (such as plague, *Yersinia,* or potato blight, *Phytophthora*), because of the ready availability of new individuals to infect. Famine caused by one organism (*Phytophthora*) affected the lives of a great proportion of the Irish population in the nineteenth century, displacing them from their homes and subjecting them to previously unexperienced social evolutionary pressures. A disease that has a low incidence in a sparse and scattered population can be disastrous in cities or cultivated fields of crops. The 1918 flu epidemic infected 500 million and may have killed 100 million of them. Young adults were susceptible and this, coming on top of the tragic Great War that killed a great many young men, left a generation of young women who had learned or had to learn to be self-reliant without husbands. A door was slammed on *La Belle Époque* and Edwardian Britain, and a very different Europe went forward to fight World War II. Society and natural selection had conspired and the human world became as we see it today.

This is why the bird flus of recent years and other errant viruses such as Ebola are of such concern. The last had a fatality rate of 90 per cent when it emerged. Humans and virus were both under intense selection pressure. Resistance to Ebola does occur naturally in humans and fortunately the virus does not survive well outside its host. Resistant people are the ones who will repopulate the world in the event that the virus crosses its own Rubicon and becomes more resistant to external conditions and moves out of Africa.

A pandemic, such as the plague, thus has the power to bring about a radical change in society and, by so doing, change the course of history and of human evolution. (This text was written, by the way, before COVID-19 and its impact on our modern society.) In the fourteenth century, before the plague struck, the social order was that of feudalism, a pyramid-like structure in which the lord of a manor held his land on swearing loyalty to the local baron who in turn owed service to the monarch. The lord of the manor had yeomen farmers who owed him loyalty, and they in turn had serfs, effectively slaves, who worked the land, in return for a living for themselves and their families. In times of war everyone was expected to answer the call of the monarch or pay a fine. This system was sustained by the farm work of the lowest echelons and no serf or yeoman could change masters to go looking for a better position. The plague changed everything. Labour became scarce and whereas before a serf would have been returned to his lord if he started looking for work, now he was welcomed with open arms. With freedom of movement, feudalism collapsed, giving way to the agrarian revolution of the fifteenth century and eventually the Enlightenment.

By the beginning of the nineteenth century there were more humans on the move than ever before in history. The British, Spanish, French, German and Portuguese sought to build Empires. The result is that, in a country like Australia that encouraged immigration, over half the population now has a language other than English spoken at home. The diaspora of the Irish and the Chinese in the nineteenth century showed that people were more mobile, with whole clans moving by boat, by roads and most recently by air. There has been a major mixing of genes, traditions of child-rearing, and social structures, from democracies to dictatorships.

At the time of writing this, the world is in the grip of a pandemic due to COVID-19, a disease caused by a new coronavirus. With 'social distancing' and 'lockdown' quarantine measures, habits of generations are being disrupted all over the world. There is much speculation that the world will never be the same again. At the very least, 'working at home' is likely to become more common; likewise with 'distance teaching'.

It is possible for every person to be influenced directly by the cultures of seven generations: those of their own, those of their parents and children, their grandparents and grandchildren, and their great-grandparents and great-grandchildren. As generations overlap, cultural communication can occur. As an example, one of us has memories of many songs of the Victorian Music Halls from his grandmother who learned them from her own mother. Similarly, it is likely that aspects of his own twentieth-century acculturation will be passed to his yet-to-be-born great-grandchildren.

Another instance of social selection, very different from what had gone before, occurred within the technical environments created by the Victorian Industrial Revolution. The great landowners seized on technology as a means to feather their own nests at the expense of the agricultural working class. They imposed a version of their existing aristocratic hierarchical society on the new culture that began to emerge. It was this elitist setting that nurtured the socially based reorientation of Darwin's work as 'survival of the fittest/wealthiest' almost as soon as the *Origin* was published. God was an Englishman and all was right with the world. To paraphrase Donald Horne (1970), only Englishmen with such a world view could sing of England in the words *of Land of Hope and Glory:*

> God, who made thee mighty,
> Make thee mightier yet!

At the same time, contemporary records show clearly that Darwin himself – and his defenders and myriad collaborators of all classes – were entirely aware of the collective thinking that led him to build his knowledge from all available sources and thereby pursue his idea of evolution by natural selection.

An even more surprising contributor to the evolutionary process is the influence of the digital revolution. This is so recent that we do not yet know the effects on populations or individuals, except that they are going to be massive. The designer of the first personal computer, and the first human–electronic partnership was Norbert Wiener (1988). He created the first human–electronic arm – only a few steps away from the science fiction cyborg. He described the personal computer as a prosthesis of the mind, somewhat like the electronic arm. Even more important, he predicted that, for the first time in the history of the planet, the global environment would be constructed by humans, and that evolutionary pressures would derive from that human-made context, a circular effect.

In his seminal book, Wiener points out that, as yet, there is no ethical system to guide us in this new world. Wikileaks, social media such as Facebook and Twitter, fake news and alternative facts are part of this story. Our understanding of evolution needs to encompass the complexity, and the essential contribution of a collective understanding to the genetic, epigenetic, cultural and symbolic dimensions of evolutionary change. Whereas in social animals, such as ants, symbols tend to be olfactory and tactile – where scent and touch have meaning – human symbols may be auditory, visual, tactile and, to a lesser extent, olfactory. Immensely complex combinations are possible, providing the glue for complex societies. A collective understanding of a collective process ought to create a better world for humans to live in. It is, potentially, niche improvement on a grand scale and, at best, will contribute to human survival, wherever on Earth humans find themselves. Societies may themselves form a culture and contain lesser cultures within. Roman civilisation, in its heyday, embraced people black, white and brindled, together with the gods of their choice.

12

OUR CLOSEST COUSINS

In which we try to identify the things that make us human.

We can now attempt to identify the origins of the characteristics that make us human – adaptability, tool use, intelligence, our social systems and, above all, our capacity for empathy. There are many versions of what it means to be human; what follows is our personal view. We argue that the characteristics of humanity do not stand outside of our biology but are part of it. Perhaps our intelligence is, for good or ill, the latest link in the golden network of Gaia. And we consider ourselves to be both ethical and moral beings.

> When the Himalayan peasant meets the he-bear in his pride
> He shouts to scare the monster, who will often turn aside.
> But the she-bear thus accosted
> Rends the peasant tooth and nail.
> For the female of the species is more deadly than the male.

In this, the first of 13 stanzas, Rudyard Kipling sets the scene for a discussion of the difference between ethics and morality. The way it is expressed will not be regarded as politically correct today, but his theme is the imperative of maternal care. The she-bear is responding to the paramount requirement of her biology in protecting her young; the he-bear has other things on his mind.

Ethics, according to the *New Shorter Oxford English Dictionary* is:

> a set of moral principles, especially those of a specified religion, schools of thought, etc.

Morals, on the other hand, are guides to behaviour:

> of or pertaining to human behaviour considered good or bad, or pertaining to the distinction between right or wrong …

We shall adopt these definitions. While we concede that ethics vs morality is a distinction that is subject to much debate by modern philosophers, it suits our purpose to use it here.

From the definitions, it will at once be apparent that 'morality' is a personal thing, coming from within. Ethics on the other hand, is a set of moral principles imposed on the person from without. Those who impose a particular system of ethics on others are inclined to think of them as absolute, entire – even fundamental – whereas morals differ according to the ways of thought of the people who are holding them and the societies they belong to. Using these definitions, a person holding an ethical principle may be completely amoral (in our view); perhaps a fifteenth-century Spanish inquisitor twisting the thumb-screws on a nonbeliever, acting in accordance with a religious ethic. Alternatively, a generally accepted moral act may be rejected in order to maintain an individual's personal moral integrity. This is the dilemma of one who conscientiously objects to killing in times of war. The ethic, the Commandment 'thou shalt not kill' must be suspended in favour of the perverted morality of jingoism.

Ethical systems are the province of human beings, and form part of the glue that holds social systems together. Mix in a bit of faith (Stephen Decatur's 'my country, right or wrong', for example) and continue to repeat the message of the Bellman in Lewis Carroll's *Hunting of the Snark*: 'What I tell you three times is true'. Reiteration is a powerful force. As long as people do not ask too many questions, you have the beginnings of a superstition. It is largely the conflict between ethics and morality that leads to the common phenomenon of cognitive dissonance, the ability to hold two or more contrasting views simultaneously.

An Australian magpie of our acquaintance has taken the first step towards superstition. It has learned that it will probably get a morsel of food if it knocks on the window. Many birds do that, but this one hops down, waits until you open the door and then rapidly turns around on the spot. The number of turns, up to four, is a rough measure of its eagerness and appetite. It has been doing this for several years now, a behaviour that probably occurred accidentally on first acquaintance is now considered essential by the bird. It is in the position of the person who is an unfortunate performer of ritual behaviour to ensure a favourable outcome in a specific situation. Professional sportspersons often show this behaviour – footballers who insist on wearing the same socks for every game, cricketers who, when batting, perform a specific sequence of actions adjusting their armour before receiving the next delivery. It is not a great step from this to communities lighting ritual fires to ensure the Sun returns at the end of the winter. Such people – and, presumably the magpie – have an imaginary tiger by the tail. They dare not let go.

Consider the canonical Ten Commandments in Exodus 20 (King James Version). The Commandments were primarily designed for the observance of the 'Tribe' and could be relaxed in the event of an external threat. The same source in another location lists another 300 proscriptions, about such things as eating unclean meat but we will not consider them here.

The Ten Commandments fall easily into two groups, each with a different intention. The first group is concerned with maintaining a proper observance of religion. They are rules that demand the acknowledgement of the power of the priesthood, on a daily basis.

1. Thou shalt have no other gods before me.
2. Thou shalt not make unto thee any graven image …
3. Thou shalt not take the name of the Lord thy God in vain …
4. Remember the sabbath day, to keep it holy …

The priests are the ones who are privileged to know God's will in this regard and thus ensure that the community shares periods of common devotion. Everyone would surely agree that Commandments 1–4 are imposed on the population from without. They are the commandments of the priests of a jealous god, imposing Jehovah's control on a Neolithic tribe from the Middle East. They are intended to focus the attention of the people on the Temple and to give them a sense of belonging to a particular community with a common ethical system.

The second group of Commandments is all about emphasising the social glue that holds a community together. Although they appear in the Bible, they are secular commandments that reflect the social issues of any community.

5. Honour thy father and thy mother …
6. Thou shalt not kill.
7. Thou shalt not commit adultery.
8. Thou shalt not steal.
9. Thou shalt not bear false witness against thy neighbour.
10. Thou shalt not covet thy neighbour's house … wife … manservant, nor his maidservant, nor his ox, nor his ass, nor any thing that is thy neighbour's.

It seems probable that Commandments 5–10 may have a far more ancient, perhaps even biological, origin. They offer a system of social morality. Commandment 5 serves two purposes. 'Honour your father and mother' suggests that moral persons should assist their parents as they age, as they owe both a debt of gratitude (a personal morality) and a responsibility to care for old persons as they are repositories of tribal knowledge (a community morality). But there is a darker side to this, the ancient avoidance of incest in all cultures as revealed in Leviticus 18:

> You shall not uncover the nakedness of your father, which is the nakedness of your mother; she is your mother, you shall not uncover her nakedness.
>
> You shall not uncover the nakedness of your father's wife it is your father's nakedness.
>
> You shall not uncover the nakedness of your sister, your father's daughter or your mother's daughter …

And so on.

This is a very explicit warning against incest with the text attempting to cover all eventualities. Such cultural taboos prevent most humans, except royal families, from inbreeding. There is a very clear biological imperative here. It is well established that continual inbreeding in humans and animals results in the accumulation of deleterious mutations. What is less well known is that numerous plants also go to considerable trouble to avoid inbreeding. Many reject their own pollen as well as that of close relatives, an exclusion achieved by a complicated enzyme mechanism. There are too many different mechanisms to be listed here, but they all testify to the evolutionary disadvantages that can be consequent on sex with close relatives.

Many animals also go to considerable lengths to avoid inbreeding. Mice can detect relatives by smell and refrain from mating with them. It is a phenomenon even observed in crickets (Simmons et al. 2006) although the mechanism is unknown. It is not surprising, therefore, that an evolutionary feature as fundamental and widespread as this should come to be reinforced by a system of morality and by folktale. The story of Oedipus Rex is a good example, so good in fact that an 'Oedipus complex' has entered the mythology of modern psychiatry and has even been lampooned in a song by Tom Lehrer.

In Sophocles's play of the same name, Oedipus unknowingly, because of a mix-up of babies at birth, had sex with his mother, Jocasta, who was also his queen. When he discovered this, Oedipus considered himself cursed, blinded himself and went to live in exile. In many royal dynasties, however, brother/sister matings are quite common as a means of legitimising succession. The possibility of substituting one baby for another at the time of birth has spawned many folk stories of changelings. Less commonly father/daughter and even mother/son pairings occur. Van Den Berghe and Mesher (1980) list 12 societies that have condoned royal incest: nine in Africa, including ancient Egypt; Thailand, Hawaii and the Incas of Peru. Others have a system of 'sister's son' inheritance on the grounds that while there might be uncertainty about the father, one usually knows who the mother is, as there are witnesses to the birth.

Commandment 6, thou shalt not kill, really needs a rider: 'except where otherwise expedient'. Thus, highly ethical and moral people will go to war to kill other highly ethical and moral people, both sides claiming the special interest of their particular deity. Often it is the same deity. Murder is common in most modern human societies, but rarely is there murder within animals of the same species. Death may occur as an accidental by-product – as in the case of rutting deer when the object is merely to vanquish the opponent, or female spiders and mantises consuming their postcoital partners rather than waste the resource they represent. This could be described as collateral damage, as humans describe the slaughter when the inhabitants of a city are 'inadvertently' incinerated by an air attack. Rarely, however, murder in response to a genetic imperative does occur. A male lion, having fought for and won a group of females, may kill off the cubs, a mechanism that ensures his own genes are passed on, not those of the dominant male of the former pride. At least, that is a usual explanation. Interspecies killing is more common – after all, what else can a carnivore do? And that includes carnivorous plants that live on poor soils and are desperate for nitrogen. Herbivores on the whole are prepared to let live and get on with living. It is difficult to think of examples of killing except incidentally in the plant world, although the strangler fig comes to mind.

Commandment 7, prohibiting adultery, at best is a pious hope, an ethic imposed on a community. The real commandment in many human societies seems to be 'thou shall not be found out'. In Western society simultaneous adultery is common but frowned upon, while sequential adultery is accommodated if not wholly approved by the ethical system. In the end it comes down to personal morality. In the animal world it is so common that there is a name for it, kleptogamy or, more vulgarly, the 'sneaky fucker strategy'. It is particularly common among species that keep harems. While the alpha male is otherwise occupied, the betas sneak around doing the best they can – which is often rather well, as genetic analyses of offspring show.

The biosphere is full of thieves (number 8: 'thou shalt not steal') from fleas, flukes and cuckoos to tapeworms. The flea steals blood and may leave something mortal behind in the form of plague. Liver flukes steal the life of a sheep and the livelihood of the farmer. The cuckoo steals the livelihood of the chicks whose place it has taken. The tapeworm steals sustenance from its host. All parasites are thieves, and some who cause the deaths of their host to complete their life cycle steal everything.

In human societies, thieving is well down on the list of imperatives – perhaps in recognition of its inevitability. Interspecific stealing – blue tits and milk, European magpies and jewellery, foxes and geese, mice and cheese to name some European cases that have entered mythology and literature – is common. In some human communities the *official* thief is recognised with exasperation but is tolerated because they are jealous of what they consider to be their preserve and discourage others. In a crisis, one can always round up the usual suspect. In literature the thief takes on an almost jovial aspect – when caught with their bag of swag: 'it's a fair cop, guv'nor' and goes quietly. There are many well-loved fictional gentlemen thieves. Robin Hood, stealing from the rich to feed the poor, is an archetype of the gentleman thief that has resonated in Britain for 600 years.

Commandment 9 could be rephrased as 'you must not tell lies'. Much of biology is beset with 'lies', with plants and animals pretending they are not there, that they are something else, that they are dangerous, that they are poisonous, that they are fitter (in an evolutionary sense). If you doubt this, consider orchids whose very existence depends on convincing their pollinators they are something else, or cactuses that pretend to be rocks in the desert, carnivorous plants, insects that mimic wasps so as not to be eaten, myriad examples of cryptic colouration. Lies, like stories, are really imagined futures.

Finally, little need be said about covetousness (10) which is largely an ethical commandment. Although the Leviticus interpretation includes human beings, the context suggests that such humans are to be considered as the property of the dominant male and the issues so raised have been dealt with in the considerations of the other Commandments.

These examples have been derived from the Judaeo-Christian oeuvre. The Mosaic Law is also observed by Muslims. The other great religions have similar strictures; thus the first four Buddhist commandments are: 'do not destroy life', 'do not take what is not given you', 'do not commit adultery', and 'tell no lies and deceive no-one'. Hindus are urged to be truthful, be nonviolent, refrain from adultery, not be covetous and be honest. These seem to be the desirable objectives that maintain social stability in human communities across the world.

The important thing now is to ask whether these attributes, either in whole or in part, can be identified in other animals. If there is no evidence of such phenomena, then *Homo sapiens* is truly unique. The evidence, however, seems to suggest that we are merely the first among equals. The evidence is all around us. It concerns consciousness and empathy and the roots of these are to be seen in animals that are evolutionarily remote from us.

A study of the palaeontology of birds (Moore and Varricchio 2016) suggests that there was strong selection for parental care. Those evolutionary lines that did not invest in it as highly as modern birds, with which we are familiar, died out. How did this come about? It apparently depended on the sort of breeders they were. There is a mathematical expression, in vogue in the 1970s and 80s, which attempted to define the rate of growth of a population (Pianka 1970). There are two important quantities; r is the growth rate of an organism and K is the carrying capacity of its environment. Ecologists described species as either *r*-selected, maximising their reproductive capacity, or *K*-selected, investing heavily in few offspring. Corals, with their copious annual egg and sperm production belong in the former group, humans in the latter.

Modern birds are *K*-selected. They make a great investment in providing safe nesting sites and weaving protective nests for their young. They incubate their eggs for weeks and, often, both male and female take turns. Where the female has sole responsibility for incubation, as in the eclectic parrot, the male feeds the female. When the young hatch, they have to be fed and protected. This requires intelligent, purposive behaviour from the parents and – dare we say it – a form of empathy. Empathy is needed to recognise danger to the young and to interpret their physiological needs, like hunger, and then to behave accordingly.

It is thought, then, that the relatives of modern birds died out because, the fossil record suggests, they did not invest heavily enough in parental care. A little examination of your own experience should convince you of the incredible bravery and aggression of modern birds defending their nests and young. Almost everyone must have seen small birds mobbing a hawk several times their size to drive it away from their nesting sites. In Australia, 'magpie season' occurs when these otherwise friendly and harmless birds swoop passers-by with a loud clack of the beak, perhaps drawing blood with their claws as they pass. Even the most stout-hearted pedestrian is wise to choose another path. Or imagine the shock, when walking across a large grassy paddock, of suddenly spying a spur-winged plover heading towards you, at your eye level, at full speed and with outstretched spurred wings as it aims to drive you away from its young that are hidden somewhere in the grass. (You have two options – run or drop flat. The second is better if the plover is close – it buys you a little more time to run for it!)

Mammals generally are *K*-selected. Kipling's Himalayan she-bear with cubs is concerned for their safety, which implies that she recognises a threat to their wellbeing. Her instinct is to attack, but she can override that, depending on how the encounter with a potential enemy develops. Perhaps a threat display will serve. Attack is a last resort when that fails.

The business of raising young over a relatively long period requires a bigger brain with the capacity to perceive a potential danger to offspring and respond as if it were a danger to itself. This is a form of empathy possessed by both sexes of many species of birds and mainly by the females among mammals. Humans have the largest brains and the longest period of care for children of any animal except elephants. In both cases, the offspring hang around until adulthood – and beyond. It is hard not to conclude that the empathic faculty is particularly well developed under these circumstances and the elephant and human ways of life provide the opportunities for it to be displayed in dealings with adults beyond the immediate family.

If one is to look for an evolutionary origin of human social morality one would be hard put to go to the elephant, however legendary its intellectual prowess, as our evolutionary trajectories have diverged for more than 40 million years. Instead, one must look at our nearest relatives, who share about 99 per cent of our DNA (Diamond 1992). We must look at the chimpanzees.

A study of chimpanzee mitochondrial DNA suggests that their evolutionary pathway and that of humans diverged 6 or 7 million years ago. Bonobos (*Pan paniscus*) and the chimps (*Pan troglodytes*) parted company some 4 million years later and are now separated by the Congo River. Their DNAs are very similar to each other, at the level of 99 per cent, but their ways of life are very different. Bonobo body proportions are similar to the australopithecines, especially apparent when they stand or walk upright.

Unlike the chimps, bonobos live in extended mixed communities of females, males and offspring. Apparently, mother–son and female–female bonds are paramount. Theirs is a matriarchal community where females exert social dominance over the males. The social groups occupy specific territories, but territories may overlap with others. Between-group mating occurs. In contrast, the male chimps are 50 per cent larger than females and run a masculine society with an alpha male. Male–male bonding is paramount. They have specific territories, each with a 'border force', and tend to avoid neighbours.

Bonobos use sex as the social glue. They engage in frequent sex in every partner combination, although there is some indication that they avoid incest. A female gives birth to a single infant every five or six years. Thus there seems to be a distinction between sex for social purposes and for reproduction, at the physiological level at least.

Fruit and leaves form the main diet of both bonobos and chimps but the latter also eat meat when they can get it. Bonobos do eat invertebrates and the occasional small vertebrate, but their diet does not include much animal protein. Unlike chimps, they have not been observed to hunt monkeys.

It is fascinating to compare ourselves with bonobos and chimps. Sapiens partakes of a bit of both dishes. We are mildly sexually dimorphic; men are rather larger than women. We live in mixed communities of females, males and offspring. Female–male and same-sex bondings are common. Alpha males, usually self-appointed, also commonly occur. Parent–offspring relationships are strong for both sexes but are particularly so for females. It is surmised that human societies were primitively matriarchal, but today both matriarchal and patriarchal systems occur, the latter more commonly. Humans occupy specific territories and guard the borders fiercely in times of danger. They show varying degrees of xenophobia, from none to extreme; the latter is especially evident when there is a war on. In spite of this, between-group matings are common. Like the bonobos, they have a strong social system but separation into recreational and reproductive sex is now achieved by artificial means. Unlike the bonobos, human society is only mildly held together by sex; religion or other ethical systems play a stronger role. Sex can be divisive; sexual jealousy is a strong motive for murder.

Bonobos have neither ethics nor morals, as we understand them; they have social behaviour. Such behaviour is not imposed from outside but comes from within, from emotional states that are a response to the immediate circumstances of the social group. The appearance of these emotional states and the appropriateness of the behaviour consequent on them has been shaped by millions of years, according to the following criteria. Do they in any way increase the capacity for positive social interaction between individuals, thus contributing to the evolutionary fitness of the individual and, through the individual's behaviour, of the group?

First place among these emotions must be held by empathy. Maternal care is probably the starting place. Female guinea pigs are not noted for their maternal qualities, as their pups, when born, hit the ground running, in an advanced state of development. They feed their young, however, for perhaps three weeks, depending on whether they have one pup or four. With only two teats, the mothers must ration their resources.

They respond more strongly to a calling pup if it is one of four than if it is a singleton. This implies, even in guinea pigs, a sense of empathy, of the need to respond to her offspring, but without comprehension.

Empathy, then, is a process of recognising the need, then matching emotional states with the 'empathee', resulting in concern for the other and, ideally, providing a helpful response. If a response is seen to be frequently successful, in restoring harmony within the group, say, it will be learned and adopted by others. If the outcome is favourable for the individual as well as the group, and contributes to his or her survival, the empathic response can then be manipulated by the evolutionary process and expanded to other situations – becoming worried about the plight of others, adopting their viewpoint and taking action to relieve the perceived distress. The decision to help does not necessarily depend on rational evaluation. This is, at its evolutionary beginnings, intelligent behaviour without consciousness of a possible outcome. This in turn can lead to altruistic behaviour, a topic that has caused many a battle between evolutionary biologists. We will leave them to it.

In bonobos, and to a lesser extent in chimps, the consolation of the victim of accident or aggression is a predictable outcome. For the descriptions of bonobo social behaviour that follow, the authors are indebted to the excellent book by Frans de Waal (2014), *The Bonobo and the Atheist*.

A bonobo, a victim of aggression who had just escaped from a life-threatening situation, sits alone, distressed and panting, licking an injury. Another bonobo who witnessed the encounter goes over to hug the victim, and to lick the injury as well. The victim begins to cheer up.

Bonobos may suffer accidental grooming injuries or get into the occasional fight. If one gets bitten, the biter always shows remorse and licks the wound and later, when the wound has healed, will inspect the wound and continue to show remorse.

To illustrate this behaviour, Franz de Waal tells a remarkable story about a male bonobo who was very protective of ageing females. In the morning he would lead a blind and deaf female out of the sleeping enclosure to her favourite spot in the sun. In the evening, he would lead her back to the sleeping quarters. One day this male bit the hand of a woman who was handing out vitamin pills and, in pulling away, he crunched down on a finger and, we must assume inadvertently, bit it off. A few days later, the woman returned and showed him her bandaged hand. The male took one look and, obviously distressed, retired to the furthest corner of the enclosure. Fifteen years later the woman returned to visit. The male immediately approached her, anxious to inspect her hand.

Recently, in bonobos, a very remarkable behaviour has been observed. It is the behaviour of female bonobos when one of their number is about to give birth. They cluster around the expectant mother, effectively excluding the males, though it has to be said that the males show little or no interest. The females protect the mother and soothe her with nurturing gestures and even appear ready to help manually. Chimps show no such behaviours, so it seems reasonable to assume that the last common ancestor of bonobos and humans also showed this behaviour, some millions of years ago.

Females generally are more likely to offer consolation than males, and friends and relatives are more likely to do so than strangers. What the bonobos displayed was behaviour appropriate to the situation. That this is not accidental is shown by an experiment with chimps. Two chimps sit side by side in transparent cages. In front of each of them is a container containing a reward that can only be got by using a special rake. Chimp 1 has an assortment of tools. Chimp 2 has no tools at all. Chimp 1 chooses a suitable tool, retrieves the reward and then pushes the useful tool through a gap in the partition for Chimp 2 to use. If the experiment is repeated so that Chimp 1 cannot see into Chimp 2's enclosure to know how the reward is presented, it will select tools at random to pass through, presumably on the assumption that eventually one will suit.

According to Scherer (2005), 'Emotions are an intelligent interface that mediates between input and output on the basis of what is most important to the organisms at a particular time'. In other words, they are part of an internal state that demands instant response. If you doubt this, remember falling in love for the first time. It is an all-encompassing feeling that demands an immediate behaviour. It is such an urgent response that it transcends rationality, and only a deeply planted ethic and a personal morality prevents the forlorn lover (of either sex) from acting on it at once!

There are very clearly elements of chimp, and especially bonobo, behaviour that are similar to human behaviour. If these were physical or physiological characteristics, for example in three separate related whale species, we would have no difficulty in postulating that they were also present in their common ancestor. Why, then, should we have difficulty in supposing that behavioural traits, which are based on an evolutionary imperative and which we admire so much in ourselves, are similar, so far as we can tell, in our hominin cousins? And why should we have difficulty believing that they had their origins in the last common ancestor perhaps 4 million years ago? Occam's Razor demands that we should not multiply our unknowns. The less economical explanation, that these behaviours arose three times quite independently in Sapiens, Paniscus and Troglodytes, is not impossible but very unlikely and must be rejected until evidence to the contrary is found. The conclusion must be that if three related species, with more than 98 per cent genetic congruity, act in similar ways in similar circumstances, then the mental processes that give rise to these similar behaviours are also likely to be similar.

For primate and human children there are two great reinforcers; the first is the ability to empathise with their siblings and others of a comparable age; the other is a desire for good interaction with adult members of the social group. The threat of physical consequences from the grown-ups promotes what de Waal calls 'one on one morality'. This enables individuals with disparate abilities and physical strength to form congenial social structures, males and females, adults and juveniles, to make a successfully functioning social group. In human history, this is first the nuclear family of hunters and gatherers, then the extended one of uncles and aunts and cousins, as available resources permit, and finally, the tribal village of several families mutually dependent on agriculture.

As part of Gaia, humans have never lost contact with all other forms of life with which, after all this time, we still share important genes and to which we are clearly related. Surely we are permitted to echo Charles Darwin's great appeal to his fellow humans: to see that there is grandeur in this view of life.

13

GLIMPSES OF THE FUTURE

In which we look forward, with some trepidation, to an imagined world in which today's discoveries have been put into practice, and consider the impact that today's social media might have upon human connectedness!

Humanity still evolves. Natural selection is part of our destiny but another layer of evolution, social evolution, seems destined to play a more immediate part. Our environment is changing rapidly; in one long lifetime we have moved from pen and ink to interactive social media. Computer intelligence and scientific discovery are shaping our future. Science fiction gives us a glimpse of what it could be like.

How will we evolve, with a world population of 8 billion individuals busily mixing genes and a generation time of 20 years? If we think solely in Darwinian terms, the answer is: slowly. The mathematics of Darwinian evolution is such that it tends to stabilise large populations, by weeding out mutations, and proceeds most quickly in marginal or isolated groups.

As a part of Gaia, humanity and its affairs is not exempt from the fate of this collective whole. Recently, we have perceived that we are having a measurable impact on Gaia. Whether for good or ill, our future depends on the continued existence of Gaia. It is therefore worrying that our impact has been brought about by tampering with one of the fundamental components of energy flow through the biosphere – that is, letting loose our stored energy, fossil fuel, all at once, in geological terms.

Already, in only 250 years since the Industrial Revolution, we have exploited fossil fuels to the extent that a carbon dioxide greenhouse effect is warming the planet and disrupting the climate. This is creating a new scenario because Gaia's regulatory systems, one of which is evolution itself, are adjusting to 'down-regulate' the effect of such changes and achieve a new steady state. Part of that down-regulation could be to curtail the activities of the causative agent – ourselves. Gaia will survive but we may not.

In the face of finite resources, the phrase 'sustainable development' is an oxymoron. The idea of evolution going on forever to greater and greater complexity cannot be sustained. As complexity increases, the amount of energy required to maintain complexity also increases. And, as we have seen in other walks of life, highly complex things are inherently less stable. Things have a tendency to fall apart and, according to the mythical Murphy, anything that can go wrong will go wrong, given enough time. Even the *Titanic* was sinkable; so was the 'invincible' HMS *Hood*, which exploded and sank in three minutes when German shells penetrated her armoury. Evolution, however, will continue to rearrange the DNA Lego bricks with not *necessarily* any net increase in complexity and anyway, what could be more complex than the human brain, with its trillions of connections? Perhaps it marks the end of our line of evolution.

We do, however, have another string in our bow, and it is social evolution. Perhaps if we can bring this to bear on our existential dilemma, we may win ourselves a reprieve. Historians are long used to thinking about human social structures as if they, the structures, were organisms. The concept of civilisations as super-organisms with limited and predictable lifespans is explored in Oswald Spengler's *Decline of the West* (1932). Indeed, by making comparisons with previous civilisations he concluded that, by the beginning of the twenty-first century, Western civilisation would be in decline as established religions lost their grip on the majority of people, and national boundaries would blur to provide overarching empires, possibly even commercial ones.

The idea that towns and cities, those that are not architect-designed like Canberra and Brasilia, arise naturally in association with natural resources and organise themselves along general principles is due to Lewis Mumford (1961). Arnold Toynbee (1972) thought in terms of the life of the city, during which it faced continual cycles of challenge by external influences, followed by response. Eventually, the senescent city is unable to meet the challenge and it falls. The three consequences of energy flow are here displayed: increased worldwide connection (Spengler), self-organisation (Mumford) and cycles (Toynbee). During these changes, humanity is continually challenged by social as well as natural evolution. Social evolution is the quicker of the two; 12,000 years ago people had to be able to live in a stone age culture and today, with more or less the same genetic make-up, they must live in a world changing under the impact of computer-assisted intelligence.

Most recently there has been interest in the organisation of human societies from researchers such as Peter Turchin (2007). Turchin was originally a population biologist who studied the way that prey and predator relationships changed. He was interested in why they cycled through boom periods of rapid prey increase followed by busts as the predator population soared then crashed as the abundance of prey decreased and food became scarce, and so on. He then realised that his mathematical description of this process could also be applied to describe the rise and fall of civilisations.

Turchin identified two sorts of cycles of civilisation. The first is a growth cycle that typically lasts two or three centuries. It starts off well enough. In the early stages, the foundation period, everyone is more or less equal but, benefitting from the improved living conditions as people cooperate in building the city, the population begins to grow. Society begins to stratify and wealthy elites form. The elites control resources and employ the less wealthy as a workforce. As the population continues to grow, the number of potential workers outstrips demand and their wages and standards of living fall.

Social unrest follows, and a destructive cycle begins. There may be revolution as social structures break down and a greater equality among the population is established. This period characteristically lasts for about two generations, when regrowth and rebuilding start. Destructive cycles may be repeated before final collapse or regeneration. From this simplified account it is clear that even human social systems are, in Darwin's words (from *Origin*),

> elaborately constructed forms, so different from each other and dependent upon each other in so complex a manner, [they] have all been produced by laws acting around us.

As well as this inherent cycling, however, there is still another tiger lurking in the evolutionary underbrush, one that Malthus and Darwin thought they had flushed out nearly 200 years ago. It is the problem of exponential growth. The human race has grown exponentially with the result that by 2030, if unchecked, it will take two planets' worth of resources to sustain us. Happily, there are signs that the rate of population growth is beginning to slow.

But the exponential tiger has other prey in sight. Just as the Victorians in the middle of the last century had failed to see the implications of exponential growth for living things until Malthus and Darwin pointed it out, so Kodak, the company that had been a household word for 100 years, fell foul of the exponential growth of information technology. Kodak failed to understand that exponential doubling affects other things than populations. In spite of having developed a digital camera as early as 1975, the company decided not to pursue the technology, concentrating instead on their stock-in-trade of now largely obsolete photographic paper. Kodak filed for bankruptcy in 2012. In 2014, after selling off assets in the form of patents it had held, they were again trading profitably but with a vastly different strategy. Kodak had fallen, one more victim to the tiger of exponential growth, this time not of living organisms but of the increase of miniaturisation of computer circuits. There will be other casualties.

transistor numbers

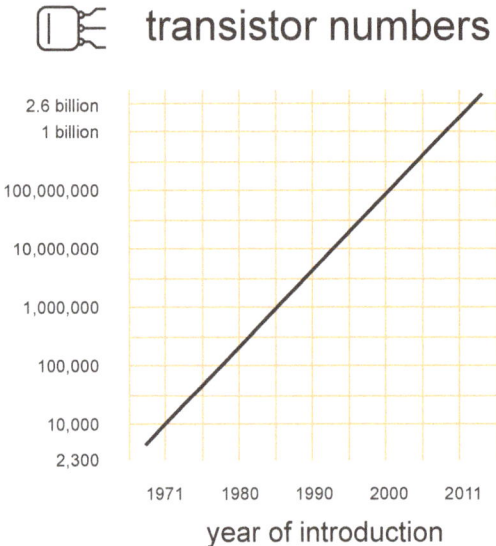

Figure 20. Transistor production.

Moore's Law, shown pictorially in Figure 20, states that 'the number of transistors in a dense integrated circuit doubles approximately every two years'. The period is now often quoted as 18 months, because modern transistors are faster. The evidence suggests that Moore's Law will continue to apply for at least another 10 or more years, as quantum computers and, more distantly, 'wetware' computers, made of neurons, already exist in rudimentary form. By that time, Ray Kurzweil (2016), Google's director of engineering, expects a computer intelligence will have passed the Turing test and achieved human levels of intelligence.

Should we be afraid? Our societies will certainly change under the impact of increasingly able computers. Do we have anything to fear physically, from attack by robots? Science fiction writers for more than 100 years have been anticipating societies in which this has happened. In particular, Isaac Asimov (1954), with his stories of dystopias in which robots play an important part, has been very influential. His three Laws of Robotics, built into robots by humans, to eliminate threat to humans, are frequently quoted. They are:

> A robot may not injure a human being or, through inaction, allow a human being to come to harm.

> A robot must obey orders given it by human beings except where such orders would conflict with the First Law.

> A robot must protect its own existence as long as such protection does not conflict with the First or Second Law.

Later, Asimov added another law, one that took priority.

> A robot may not harm humanity or, through inaction, allow humanity to be harmed.

Asimov was a writer of science fiction. His Laws of Robotics were carefully thought out, but they were literary devices, developed so he could write stories about the ways that his characters, robots or people, circumvented them. They are too vague, lacking precise definitions of 'harm', 'benefit' and 'human'. Asimov's imagined worlds consist of people not too dissimilar from you and me, with robots that are apparently at the top of the artificial intelligence tree. Louie Helm, of the Machine Intelligence Research Institute, has blogged that even if Asimov's vision came about, it would be transient as humans begin to deal with the ethical problems of a machine super-intelligence that transcends the robot era.

But suppose a robot decided that it was in humanity's best interest in the long run to reduce the population. It is already conflicted by the first and the fourth Laws. If it decides against killing the 'surplus' population it must resort to other means and disregard the Laws. Indeed, one new scenario has recently been reported in which the Laws may be circumvented. Apparently, self-driving cars can be fitted with an 'ethical knob' to override the Laws which, if pressed by the passenger, allows the robotic 'driver' a choice when about to be involved in an unavoidable accident. It may seek the least harmful alternative, and allow its the passenger to be killed if it will bring about fewer deaths than taking evasive action.

Our capacity to read the future depends, in part, on how well we can understand the past. In the *Phenomenon of Man* (1959), Teilhard de Chardin saw a trend in the increasing evolutionary complexity of nervous systems and brains and conceived a future when all humans were connected in a planetary hook-up that he called the noosphere. His idea of a noosphere was compounded of his religious beliefs and a sort of telepathy. Writing 70 years ago, he had completely failed to see the rise of computer intelligence and the technology that permits the construction of a 'mechanical noosphere' that is embodied in modern social media. He was, however, certainly correct in his view that increased connectedness between individuals would change the world, but not in a way that he expected. Isaac Asimov, however, could conceive of such a future, where robotic intelligence was commonplace. In *Caves of Steel*, written about the same time, Asimov describes a society so enamoured of its capacity for computer communication that its individuals have come to live in splendid and sterile isolation, communicating by the equivalent of Facebook. It is the extreme of the tendency towards personal physical privacy that has been developing in the West since the eighteenth century.

If we want to look into the future in the relatively short term – say, 200 or 300 years from now – it may be good, first, to look back the same distance to take note of trends. The present book is written by a couple of dinosaurs, children of the 1930s, so we have had the opportunity to notice more than most. At school, we were all taught about the Industrial Revolution and how a set of technologies based on advances in iron smelting and metallurgy, made possible by the availability of abundant coal, changed the face of Britain. The industrial future was, and still is, embodied in the Iron Bridge that crosses the River Severn. It was made of cast iron and opened in 1781. This was the first triumph of a modern disruptive technology, a term used by Clayton Christensen (2003) in his book *The Innovator's Dilemma*.

A disruptive technology is a game changer. It can be large, like the Iron Bridge, or the Model T Ford that did away with carriage horses. It can be the accumulation of small discoveries. One of us, growing up in Australia in the 1930s, remembers the iceman. His job was to patrol the streets with a horse-drawn cart that had an insulated compartment containing large blocks of ice. These he sold in chunks to the local householders, while small children clamoured for the ice chips. This was a livelihood that was doomed when the first powered domestic refrigerator appeared. It was also one less task for a carthorse. There are many, many other examples. Indoor flush toilets did away with the night-soil collector and his poor old horse. The streets used to be loud with the calls of the rag and bone man and his horse-drawn cart. In England, the Italian ice-cream vendor on his modified tricycle, the itinerant Spanish onion seller on his bicycle, his strings of onions draped over his shoulders, the dustmen collecting dustbins, the coalmen, also with horse-drawn carts, delivering coal and coke, in hundredweight sacks poured straight into the cellar, are gone. When the word processor came on the scene, the typewriter was doomed. Who now remembers logarithm books, slide rules or mechanical calculators? Who now remembers typing on wax sheets for the Gestetner duplicator to make copies? Who remembers what the world was like before computers and Twitter? We have been inundated by a deluge of new technologies for the last quarter of a century and have adapted to them. But new technologies make new societies. The discovery of fire, the knapping of flint, the smelting of metals, the first labour-saving machines, the first motor cars all changed the societies in which they first appeared. Here are some new technologies that may have profound effects in the future. Most of them are discussed in more detail by Fagan (2017).

1. Social media. There will be a higher level of computer-driven connectivity but reduced physical interpersonal communication. Already it seems that dating websites are liberalising humanity by increasing the number and type of people available outside the usually local group of potential mates.

2. 3D printing is becoming better and cheaper. It says much about the human race that one of the first 'printed' things to receive public attention was a ceramic handgun. As 3D printers become even

more common and accessible, only the easily transported printing programs, and access to a local machine, will be needed to build a wide range of products. (Since this paragraph was written several months ago it is already out of date. Personal 3D printers are now available for a few hundred dollars.)

3. Electric self-driving cars will lead to far fewer cars on the road: fewer on-road parking spaces and fewer household garages will be needed, as cars will be effectively 'parked' while travelling. Cities will be revolutionised by the freed-up space. Insurance companies will lose an important part of their business. The carnage on the roads will be vastly reduced, if not eliminated, freeing up many beds in hospitals.

4. Agriculture and animal husbandry will either be further automated, or, as promoted by the regenerative agriculture movement, natural systems will be allowed back into the equation. In developing countries, a surge in education levels will allow the continuation and improvement of appropriate local agricultural systems.

5. The first 'meat' grown artificially from stem cells has already been produced. It requires much less water per gram to produce than a real steak and far less space, and results in much less methane production so is less harmful to the planet. Previous attempts at artificial diets have, however, failed, due to the inability of humans of the time to identify the myriad contributions to a 'healthy diet' that needed to be included. It remains to be seen whether this ignorance can be overcome.

6. There are apps that can read facial expressions. If one can be made that can tell, with even an 80 per cent level of accuracy, that someone is lying, imagine political debates and interviews in the future where it is being used! Think about the effect it might have on personal relationships where one partner can be *shown* to be lying!

7. Uber is a car hire organisation that owns no cars; Airbnb is a hospitality organisation that owns no hotels. Their businesses are all about connectivity; putting people in touch with suppliers. There are concerns for the futures of the automotive and hotel industries, already complaining that these new collaborative arrangements evade taxes and hard-won health and safety legislation.

8. Medical diagnostics and surgical operations by robots are on the increase. Machines already perform many of the simpler procedures. The doctor's role will become largely one of reassurance and

direction to the appropriate service, the provision of counselling and perhaps providing first aid. There will be important implications for welfare systems.

9. Law is a recorded accumulation of previous human decisions that create legal precedents and address most of the ethics encased in the Bible and the Koran. In a world dominated by digital communications, it is expected to become fully programmable – and infallible!

10. Pattern recognition software will enable instant identification of faces in crowds and permit constant supervision. It will allow the identification of perpetrators of antisocial behaviour and permit constant supervision of everyday events. There have already been attempts to introduce an identity card into to Australia, so far, heartily rejected. Now, given the activities of data miners, it may be unnecessary!

11. Power will be cheap and clean. Solar energy production has been on a rising exponential curve for years. In 2017, more solar energy systems were installed worldwide than fossil fuel power stations. With cheap power comes cheap and abundant freshwater by desalination of seawater. Electrolysis of seawater will provide unlimited hydrogen as a power source and return oxygen to the atmosphere.

12. CRISPR (recall: 'clustered regularly interspaced short palindromic repeats'). We have already discussed CRISPR in Chapter 4. In short, it is a simple way to edit the genetic make-up of an organism. Small lengths of DNA are removed from, or inserted into, a pre-existing DNA molecule. The first CRISPR-edited baby has been born.

This list, which is certainly not exhaustive, concerns things that are happening now. At the very least, they will bring about a restructuring of the human workforce. How they will interact, and what that will bring in the medium term, is far from clear. There will no doubt be unforeseen, emergent consequences as a product of all this new connectivity.

It is thus at least a 12-horse race. Were we to bet on a couple of winners, they would be numbers 1 and 12, the impact of social media, and CRISPR.

Already there is evidence to show that the connectivity of social media is creating worrying consequences. Twitter, apart from being the preferred mode of communication of a former President of the United States, has had a major impact on young people. Things can be said on Twitter that former social conventions would have prevented in face-to-face conversations. This loss of reticence, combined with the power of images on Facebook, can have a devastating effect on undeveloped minds, driving some to suicide. Recent reports in the conventional communication media describe the concern that schoolteachers have for children, to the extent that, in some schools, phones are confiscated during school hours. One result of this, according to some teachers, is that breaks and lunchtimes are once again ringing with laughter and buzzing with conversation. It seems healthier than sitting around in groups, communicating with thumbs. One surprise, though, is that parents feel uncomfortable with the embargo on phones. In these days of close parenting for fear of bullies and paedophiles, parents feel intense discomfort in being cut off from their children and 'not being there' for them, even during school hours.

The full impact of CRISPR has yet to be revealed. It provides unlimited opportunities for altering the genetic composition of any organism on Earth, from the single cell to the blue whale. Food crops and domestic animals are obvious targets, with the desirable objective of improving yields and increasing resistance to disease. Once the genome is known, the strategy is to make the necessary changes in the fertilised egg, allow it to grow to maturity and to breed from it. It is possible to do this in humans, weeding out the sections of DNA that may lead to unfortunate genetic disease in the adult. In humans, it is a dangerous strategy, as so much of human DNA is silent. However, in time these problems will be overcome and the technology may move from deletion of the deleterious to the attempt to make positive changes that will be expressed in the adult.

Will we learn from the eugenics disasters of the twentieth century? Or will we attempt to make super-humans, a race of beautiful geniuses who will take over the world? Is it all a dream? Or a nightmare …?

The remainder of this chapter describes a future imagined by the authors, based on current science and inspired by the works of Fagan (2017), Asimov (1954), Aldous Huxley (1932) and Winchester (2004).

Days in the lives of ...

Part 1

Mary woke up with an intense feeling of relief. She lay still for a while, reliving a nightmare in which she had been walking along a huge indoor shopping mall in the middle of a large crowd. She cringed at the thought of the press of people. Ever since she was a little girl she had been scared of crowds. She supposed it was the fear that had been passed on to her by her parents, of contracting the Disease. Her parents, who had lived through the worst of the Great Infection years, were forever traumatised. They had impressed on her the lethal danger of crowded spaces. She gave a little shudder and came more fully awake, wondering what the time was. On cue, her implant displayed 25.31.07.2.2.08.2218. Second of August, it was her birthday and she was 50 years old.

Time to check on everyone, she thought, calling up ChatterBox in her implant. She quickly ran through her friends' previous night's exploits, sent them a Happy Tuesday message, and a private message to her life partner. Ever vigilant, her implant sensed her mood and her heightened emotion and obligingly provided 10 minutes of erotic sensation.

Totally relaxed now, Mary started to contemplate the future. She was, after all, only 50. With reasonable luck, and regular halting of the ageing process through replacement of the DNA lost from the telomeres at her chromosome ends, she could look forward to another 100 years in perfect health. Her mind turned to her eggs and the families that she had always planned. Perhaps it was time to start the first one.

By the time Mary had been born, in 2168, they were just about over, those terrible decades of haemorrhagic virus disease. The Great Infection had finally run its course. GI was a variant of Ebola virus that first emerged in West Africa in 1976. The only good thing about it was that if you survived infection you were immune. After about 75 years in which it had been confined to the African continent, it had finally entered the global population in 2050 through the worldwide Stratospheric Travelnet. Some said it was the ultimate act of terrorism, the Disease being carried by a man who, while himself immune, had been infectious to others. He had travelled widely around the globe, spreading the Disease, before being killed in a freak accident. He had an asymmetric encounter with an 80-tonne *Argentinosaurus*, regenerated from fossil DNA, in a South American game park. He was reduced to a smear on the road by one sweep of the giant tail. Once the wandering source had been eliminated, the Great Infection was gradually brought under control but not until it had done away with about half of the human race. The human race took a deep breath and counted up the dead and found it to be about 4 billion. Fortunately, the Disease had never spread to the colonies on the Moon and Mars. They were newly self-sufficient and they had quarantined themselves.

Machine intelligence had reached the stage of self-maintenance by the time the crisis came. Western society, China and Japan had each been held together by the efforts of the world of computers and robots. In the USA, the easy availability of 3D-printed firearms had led first to the collapse of central government, then to the emergence of pseudostates permanently at war, as the more populated ones strove to consume the poorer ones.

Along the way, men as sperm carriers had become collateral damage. It had been known as early as 1950s that sperm quality was declining. The cause was unknown, although the rising concentration of endocrine disruptors used in human and animal population control, and now to be found as pollutants in drinking water, were suspects.

A remedy for sterility only emerged with the growth of stem cell technology. Stem cells were harvested from bone marrow and maintained in special nutrient culture medium. Tweaking them, by careful adjustment of the composition of the nutrient medium, encouraged them to develop along different pathways, becoming blood cells, muscle cells and those of other types, including sperm.

Stem cells from the healthiest males were therefore stockpiled in special sterile centres that had been established and staffed by medibots. As the cells matured, the DNA of the sperm was checked and those conforming to the best genetic standard, after CRISPRing, were frozen for future use. On the presentation of a CRISPR certificate, any woman could make a withdrawal, submit her eggs for fertilisation and bank them in suspended animation until required. They would be released to her on production of an Approval to Reproduce.

Mary had had her Approval Form for several years. Now she got it out to check the date and, with satisfaction, saw it still had a year to run before expiry. Theresa, her special friend and lover, had got hers at the same time, and they had always planned to hatch a pair of eggs together. Each month they spent a week at one or the other's house. They were both professional women so they could afford a fully self-driving vehicle and docking facilities at their homes. This allowed them to visit one another without exposing themselves to the outdoors, for Theresa, too was agoraphobic, having also learned to be terrified of people during the Great Infection.

Mary was a gifted ceramicist. Her workroom still had a potter's wheel as she liked to feel the clays and the pastes she worked with changing shape under her fingers. She made pieces of all sorts, household utensils, cups, ornaments, artworks. This laborious, repetitious but strangely satisfying side of the creative art had remained, although duplication was automatic. All she had to do now was to think of the original idea and make the item. Her implant would record the movements of her fingers, convert it to a manufacturing program and send the record to a 3D printer that would store it in its capacious memory. The item would be printed while she could continue enjoy the stimulation of the creative thought. Buyers could order any item except for those commissioned by people who had paid a healthy fee to keep their purchases unique to themselves. She also did occasional commercial work that she sent off to industrial printers for reproduction in bulk.

Theresa was also an artist, and perhaps it was this that was the foundation of the friendship. She was, with the help of her implant, a 'visualiser'. She imagined stories about the future, mainly for children, and her implant captured the images and relayed them to a storage computer. On request, the computer would transfer them to the implants of specially assigned critics for approval and editing. She too was very successful, and her imaginings were very popular and appeared on a variety of social and communication media. Her eggs had also been passed by CRISPR and stored to await future activation.

Mary's enthusiasm for parenting increased as she opened the private section of her implant and ChatterBoxed Theresa to explain her decision. Theresa was delighted, for she, too, had thought it was the right time for the first parenting episode of their lives. They agreed to meet at once.

Mary went to the docking pod to wait for the transporter. Her excitement and happiness grew on the short journey and she barely noticed the creeper-covered ruins, where once had been terraced housing, nor the abundance of bird life and flowers and the absence of livestock on the short journey. Theresa was at the dock to welcome her and they immediately went into the studio to discuss plans.

First, they checked Theresa's reproduction permits and all was in order. Both women had decided on employing exo-uterine technology, as neither wanted the bodily distortions caused by a biological pregnancy. There followed the delicious decision of choosing the model. In the end, as they were both well-off, they picked the most expensive: Bokanovsky Multi-Ectogro, the Rolls Royce of its type.

Theresa had opted for the 'full pregnancy hormone experience' because, she said, she thought it would make her a better visualiser. The health robots that already monitored Theresa's immune system could easily take care of this. Mary elected to take only the 'lactation' package both for convenience and so as not to forgo the opportunity for intimacy it would provide with the newly decanted babies.

Together, they decided to create a special menu for a celebration dinner and it was a fine banquet. The centrepiece was a roast made out of compacted muscle cells derived from the stem cells of an Aberdeen Angus (who was still grazing happily in a paddock in Scotland), assorted vegetables, followed by blue-green sweet cakes made from a flour of ancient algal cells. The main dish was accompanied by a surprising number of fermented foods to placate their intestinal biomes. There was yogurt, kefir and kimchi, fermented green beans and sauerkraut, with kombucha and cultured buttermilk to wash it down. There were different varieties of cheese to go with the cakes. All this food was accompanied by a magnificent red wine, which they had ordered for the occasion. It was quite fresh, having been formulated only that morning, but the winemaker was a genius and had crafted into it 20 years of simulated age that would please the palate of the most fastidious connoisseur. Needless to say, it wasn't cheap.

The next morning, Mary and Theresa completed the applications and submitted them for approval. In the usual computer efficient way, acknowledgement was instantaneous and the two approvals arrived two minutes later. One hour after that, two containers that looked rather like the small plastic butter containers you get in restaurants were delivered by courier. They arrived together with holographs that showed the fertilised ova from all angles for their records.

They inserted the pods into the Multi-Ectogro. Disdaining growth rate adjustment function, they set the dial at an old-fashioned nine months and pressed the button. The incubator hummed into life. In their excitement, they failed to notice the occasional flicker on one of the thermostat indicators. Nine months later they were delivered of two baby boys.

Part 2

The following is a copy of what remains of an address that was apparently mindcast in the early years of the twenty-fourth century. It was found in a time capsule buried under the body of a well-preserved cyborg, in the Cybermorgue of a Euthanasia Centre at the South Pole.

Our name is Charlie – Charles or Charlotte depending on your preference. We were born into a world where gender was considered to be a trivial attribute, compared with one's intelligence. This is reflected in the fact that we are a person who was born as an amalgam of both genders and brought up accordingly.

At the time we were born, in 2219, life expectancy was about 150 years. Since the new outbreak of GI, whose rapid spread was assisted by the deficiencies of a largely synthetic diet, much research has been devoted to providing an optimal diet for citizens. We now eat food that has been grown and created according to the prescriptions of a balanced hunter-gatherer diet. One of the effects of this has been to bring about a further significant increase in life expectancy. This has stimulated even more research, especially on an effect of stem cells that was first observed centuries before we were born. It was found that regular transfusions of stem cells into mice considerably lengthened their lives. What worked with mice also worked with humans.

When we were small, some of our stem cells were harvested and then grown in perpetual culture. Now, they are continually used to replace the ageing cells whose damaged DNA contributes to the ageing process. We could expect to live to 400 but we suspect that few will have the stamina – or be allowed – to last the course. We are now, in 2419, 200 years old, currently the socially permitted limit of our life, if one is not a double-alpha. We are therefore keen to complete this record before the Euthanasia Team arrives.

We are what used to be called a cyborg. That is to say, many of our natural functions have been replaced by prostheses. Our implant is an obvious case, as it gives us enhanced hearing and vision. Because of a malfunction with the thermal setting of the Bokanovsky Ectowomb purchased by our parents, our limbs did not develop properly after birth. As a result, we were early enthusiastic supporters of the prosthetic movement and without much misgiving we volunteered to test new prosthetic limbs that had been and were being developed. Any misgivings we might have had were solely about the discomfort that surgery might involve. We knew that the stem cell technology would, in any case, allow us to grow new limbs, at the cost of about a year's worth of time and effort. Our misgivings were baseless, however; and equipping us with artificial limbs was an enormous success, as they greatly outperform natural ones. The mini-computers in each of our prosthetic limbs amount to what is effectively a dispersed brain – somewhat analogous to the eightfold octopus brain – that communicates, via our implant, with our natural brain, so our mental acuity is much enhanced.

We won't describe the tribulations of living through the period of the evolutionary accommodation that occurred between Gaia and humans. The mean atmospheric temperature, due to the elevated carbon dioxide levels, rose about five degrees. Suffice to say that, even though burning of fossil fuel was outlawed quite early, and replaced with solar energy, there was a considerable overshoot. Temperatures continued to rise as greenhouse gases were released from the melting ice, the warming tundra and the sea and from the millions of fermentation chambers that were wild animals. The sea level rose 30 metres. Humanity, which once considered itself apart from, rather than a part of, Gaia, learned a bitter lesson. In was in this period that stem cell technology blossomed, and 'meat' was created in closed systems that recycled resources, including carbon dioxide. The only greenhouse gases are those produced by humans as they eat the product.

The change went from the turmoil of transformation to a new steady state in a century, from about 2250 to 2350. It is fortunate that we now have ways for us to send our story mind-to-mind, via our implants, so you can have the full experience of the years we are remembering.

At the beginning of the change, communication had gone not much further than wireless connection between technical machines. Human beings were just beginning to struggle with the disruption of their social and natural environments. The changes in the summer of 2300 seemed abrupt, even though they had been predicted since the mid-twentieth century. At that time technology was at its height. Shuttle planes were, like Shakespeare's Robin Goodfellow, putting a girdle round the Earth in very nearly 40 minutes. The timetable was generous. You could be picked up and put down at a thousand different places around the planet, although that didn't stop people complaining.

They should have been happy with what they had. Technology has had to be reinvented to fit our new environment, and now the supersonic people carriers that were once everywhere have had to be mothballed as the stratosphere is too turbulent. A hundred years ago, in 2319, a shield of reflective dust was injected into the stratosphere to increase the Earth's reflectivity, in the fond hope that it would save the planet from the planetary warming. This was done rather than address the problem at its root, the overuse of fossil fuels. Sadly, the shield has done its job too well. Plants have been dying and biodiversity is diminishing as the number of plant species is too rapidly being reduced. There is ice formation at the poles once again and it is probable the Earth has been tipped prematurely into another ice age. Food is becoming scarce and expensive. People once again are beginning to have recourse to the synthetic foods that have already proved disastrous during the Great Infection. There seems to be no will to tackle the problem of removing the reflective dust in the atmosphere …

(The narrative breaks off here, as apparently the Euthanasia Team arrived a couple of days earlier than scheduled.)

Well, we had fun dreaming up a couple of science fiction scenarios to accommodate many of the issues to which humans will have to adapt themselves in the future. Ignoring the cries of 'don't give up your day jobs', here is the list.

1. Stratospheric Travelnet. Space travel.
2. Pandemics (this was written well before the appearance of COVID-19). Global terrorism and the collapse of the USA.
3. Failed climate manipulation; disastrous global cooling.
4. Abolition of animal farming; unsuccessful chemical nutrient diets.
5. Cranial implants connecting directly to brain and the development of ChatterBox, a social medium that talks directly to implants, also allowing direct communication between people via implants; prostheses connected to the brain by implants.

6. Very advanced stem cell technology. Limb regeneration achieved, permitting voluntary limb amputation to take advantage of superior prostheses. Appearance of cyborgs, part human and part machine.

7. Self-maintaining artificial intelligence. Advanced medibots. Self-driving vehicles and other robots.

8. 3D printers commonplace. When we started this book they cost tens of thousands of dollars. The other day a home unit was on sale for $256.

9. Cryogenic gamete storage perfected, allowing bureaucratic control of reproduction and population, leading to the abolition of reproductive sex. Recreational sex only permitted; orgasmic centres in the brain directly stimulated by implants. Hormone levels manipulated via implants creating an intersexual humanity. Total exo-uterine technology for embryonic development and gene editing by CRISPR.

10. Anti-ageing DNA technology; girlish women at 50 years of age.

11. DNA resurrection technology – e.g. *Argentinosaurus.*

12. Four-hundred-year life expectancy and social control of euthanasia.

14

WEAVING THE
GOLDEN NET

*In which Darwin's vision leads us inevitably to the concept of Gaia,
a unified, self-organising, self-sustaining, finite, world.*

The connectedness of the biosphere is embodied in the concept of Gaia.
It was the name suggested by a great writer, William Golding, to personify
the insights of James Lovelock. Another great author, this time a woman,
AS Byatt (2012), suggested another evocative image.

> The surface of the earth was like a great embroidered cloth, or rich
> tapestry, with an intricately woven underside of connected threads.

Both images emphasise the connections that join living forms and anchor
them to the rocky mantle of the Earth and the depths of the oceans.
In space they reach the top of the atmosphere and they penetrate deep
time, back to the first cell and beyond. It is, above all, a giant integrated
whole, a self-maintaining system.

Gaia, in Greek mythology also spelled Gaea, personified Earth. It is not
the first time she has been recruited in the name of science. In fact, in
another role she has whole disciplines – geology, geography, geometry
– named after her. The Gaia of the Greeks was the ancestral mother,
the 'broad-bosomed', all-embracing nurturing entity. She is brilliantly
evoked by poet Henry Lawson, in his poem *On the Night Train*, as the
'Mother-Bush',

Have you seen the bush by moonlight, from the train, go running by?
Blackened log and stump and sapling, ghostly trees all dead and dry;
Here a patch of glassy water; there a glimpse of mystic sky?
Have you heard the still voice calling – yet so warm, and yet so cold:
'I'm the Mother-Bush that bore you! Come to me when you are old'?

Margulis's acid 'Gaia is a tough bitch!' was a response to accusations of resurrecting an outdated and fundamentalist concept of a nurturing Mother Nature, the idea illustrated by Lawson's poem. Lawson, however, is singing Gaia's impersonal qualities – 'so warm, and yet so cold'. Indeed, Gaia is tough; and unforgiving. Humanity, alone among organisms, has a choice and it is a stark one. We can be part of Gaia or we can join the 99+ per cent of all those species that ever lived and have now become extinct. In fact, it is no choice at all but a biological imperative. Our long-term future depends on acknowledging that we are part of Gaia, for Gaia is a finite entity and her premature end might be brought about either by our bungling or by cosmic events.

If Gaia's long evolutionary journey is not cut short by an errant asteroid or thoughtless global sterilisation caused by the incessant scientific curiosity and warmongering of the third chimpanzee, she might last another billion years before the expansion of the Sun boils the oceans. Going backwards, halfway into that span of time takes us to the beginning of the Cambrian and its already interesting and complex forms of life. Regressing a further half billion and we are at the start of multicellular organisms. The billion years or so that is left is thus a huge amount time for life to evolve even further. Surely it is worth making the effort to go along for the ride for as long as we can manage it.

Gaia's net is composed of the countless strands of individual genes and sets of genes that extend in time from the first cell to the present day. It includes all the interactions between the genes, chromosomes, all the ecosystems composed of many cooperating species, all the connections with the geosphere and the atmosphere, all the cycles of life, of geological and atmospheric cycles, the chemical cycles of carbon, nitrogen, oxygen and other elements. Gaia is a huge, single entity of interconnected cooperating systems that can maintain itself as long as the Sun shines. The energy from the Sun sustains the whole net; without it, the system disintegrates. Natural selection is the weaver, the process that creates more and more complex patterns as time passes.

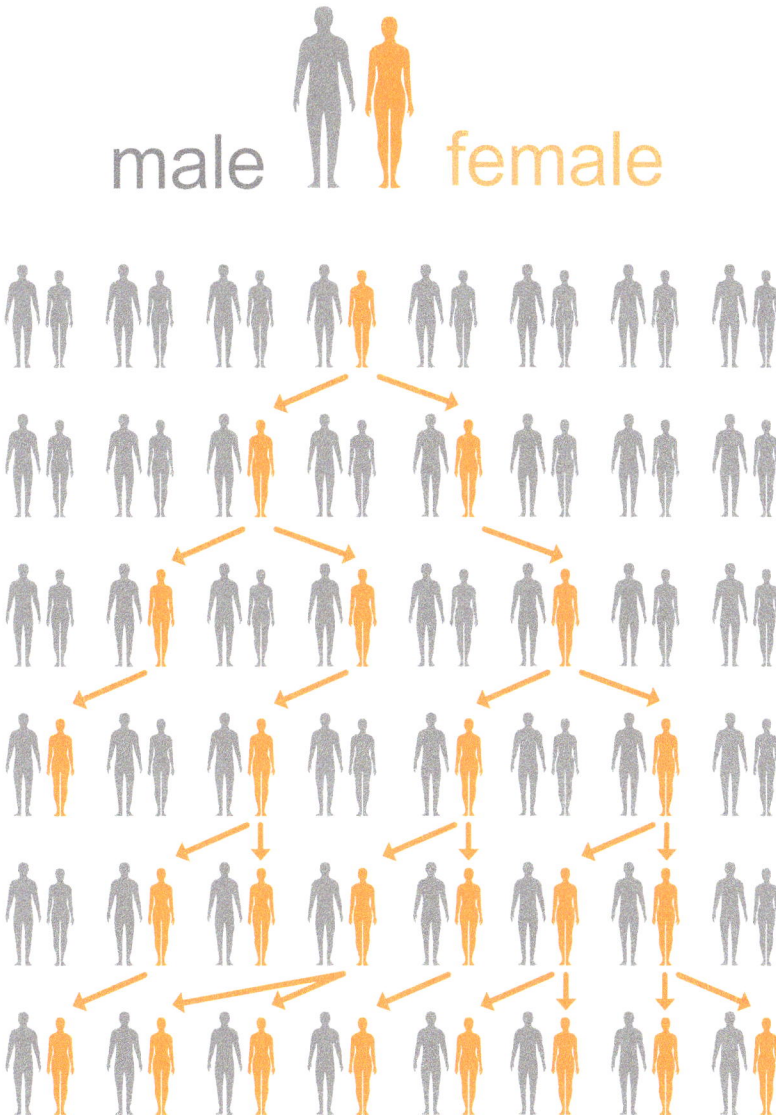

Figure 21. Mitochondrial inheritance.

We can now examine a few of the threads. Inside the cells of humans, animals and plants are found the small, sausage-like organelles that are called mitochondria, what is left of a symbiont. They form a strand in their own right, but they demonstrate some other remarkable properties as they have their own mitochondrial DNA (mDNA) and are able to divide. But there is another important characteristic.

The mDNA from these ancient modified intracellular symbionts only passes, in humans, through the female line (see Figure 21). Thus, your grandmother received her mDNA from her mother, then passed it to your mother and your mother passed it to you. If you are a woman, your children will also get yours. If you happen to be a man your children will never have yours. Instead they will have that of your partner – which was the mDNA of her mother, and her mother before her and so on. Sadly, sir, from a mitochondrial point of view, you are a dead end! But … if you have a sister, your maternal mitochondrial line goes on through her children.

In scientific literature, there is a wholly imaginary woman called Mitochondrial Eve, mEve for short. mEve is the ancestress of all living women. The figure shows how her mitochondria might spread through a population. By sampling mDNA from all over the world and identifying the range of variation in mDNA in different populations of people, and assuming a constant mutation rate, geneticists can work out approximately when this most recent ancestress of all living humans was alive (Cann et al. 1987). It is a statistical estimate, based on the rate of mutational change and is always being refined, but the current view is that she lived about 200,000 years ago. This is about the time of the great migrations of modern humans out of Africa.

A generation is usually taken to be 25 years, so mEve lived at least 8,000 generations ago. A similar study of the Y chromosome in men, not present in women, puts an ancestral yAdam at about the same generational distance (Poznik et al. 2013). Of course, this is only a statistical calculation and does not mean that yAdam and mEve knew each other – the odds against that would truly astronomical. And anyway, the couple themselves *are* imaginary!

There is an 800-year-old story about a fictional Emperor of China who wanted to reward a humble peasant who had done him a great service. Rejecting jewels and land, the peasant said he only needed sustenance and, pointing to a chess board, he asked humbly that one rice grain be placed on the first square, two on the second, four on the third and so on, doubling with each square. The total number of rice grains when the 64 squares have been covered is 18,446,744,073,709,551,615. This is many times the rice production of the whole world.

It is also the number of descendants one woman *might* have after 64 generations – assuming that every one of her descendants did her reproductive duty and had only two offspring, that there was no inbreeding or back-crossing, and that deaths from famine, disease, accident and old age did not occur. When dealing with these progressions, the numbers become eye-wateringly huge very quickly, the observation that provided Malthus with his insight and Darwin with his inspiration. Thus, Genghis Kahn was liberal with his genes and they are now carried by millions. William the Conqueror's genes are likewise to be found, probably, in all native-born English people. Her Majesty the Queen can trace one unbroken line back to William. To perform this trick, you have to know who your ancestors were!

The strands of Gaia's net are made up of the sequential generations of all life but to get a real feel for deep time it is helpful to think in very long human lifetimes, of 100 years. Think about this; when the authors were babies, they *could* have been cooed over by someone who had been alive in 1837, the year Queen Victoria ascended the throne of England. One more century back, and a newborn might have been inspired by someone who had listened to JS Bach playing his as yet unnamed Brandenburg Concertos at home. A hundred years further back still and King Charles I of England lays his head on the chopping block. Ten such lifetimes from now and people could be watching the Norman Conquest; 20, the Roman invasion of Britain. All of English modern history is therefore embraced by 20 long lifetimes, 80 generations, in evolutionary terms, a very small number.

But other living things are just as remarkable as humans. Ancient cells, such as the intestinal bacterium *E. coli*, double every 25 minutes or so; its life span is equal to its generation time. You may care to calculate the number of generations this ancient cell has passed through since its earliest ancestor and ponder on the uncountable opportunities provided for natural selection in that time!

At the other end of the scale, a strong contender for the title of the oldest living thing is a species of bristlecone pine tree about 5,000 years of age. The oldest known bristlecone pine alive today was therefore 3,000 years old when the Romans arrived in Britain. It had germinated before the first stone was erected at Stonehenge. Ten thousand years ago the last ice age ended: only two bristlecone pine lifetimes end to end. There were only a few million people worldwide and they were all hunter-gatherers.

Agriculture in Asia started about 12,000 years ago, and Eridu, the first of the Sumerian cities of Mesopotamia, was established after a further 500 years. Just two bristlecone lifetimes have therefore leapfrogged us past the Egyptian, the Greek and the Mycenaean civilisations, over the Iron and Bronze ages and into the Neolithic. Seven hundred and fifty thousand years before that, or just 150 bristlecone lifetimes end to end, marks the start of the Old Stone Age. Fossil evidence suggests that the earliest known member of the bristlecone pine species lived more than 40 million years ago. At that time, the ancestors of humans were little more than tree shrews, although, being tropical, they never set eyes on a bristlecone pine. That should give you a sense of proportion, if nothing else!

The information contained in our DNA connects us to every person, every other living thing alive. If we take the process far enough back, way beyond the current mEve and yAdam, it connects us with every hominid who ever lived. Back still further the network connects with all animals, plants and fungi until we eventually arrive at that first cell that was formed by a momentous symbiotic event.

Many scientists and science fiction writers have tried to scry their crystal balls for a glimpse of our future. Their stories are often entertaining but, realistically, these are just more 'imagined futures', the creation of which is one of the important functions of the human brain, and one that has played so large a part in our survival. One problem is that these futures are often so far away in time that we have no way of identifying and choosing the most likely one. There is time yet for lots more evolutionary threads to disappear or others to be uncovered by probing life forms of Gaia.

Humans are on the adaptive peak of only one Mount Improbable, to borrow Richard Dawkins's (1996) vivid metaphor, in an adaptive landscape where there are several ranges of improbable mountains, the peaks of which are occupied by different groups of organisms. Setting aside, for the moment, the ubiquity of modern representatives of ancient cells, the peak for absolute population numbers is probably occupied by nematode worms; that for complexity of hive behaviour by the termites, ants and bees; for animal size, by marine organisms (the blue whale at 30 metres, whale shark at 12 metres, the lion's mane jellyfish with a bell of about 3 metres and tentacles extending 40 metres); for longevity, the bristlecone pines; for the largest living organism, a fungus. It is a honey fungus, measuring 3.8 kilometres across, found in the Blue Mountains in

Oregon. Normally, one would only see the fruiting bodies (mushrooms) but they are connected beneath the ground by a vast network of interlinked branches all carrying the same genome.

And so on. It is possible to say that humans excel in intelligence, collaborative enterprises, in tool making and tool-using to support their continued existence, but at the same time unforeseen outcomes threaten to cut it short. There are others waiting their turn should humanity make a mess of it.

The effects of evolution on humans are best seen in the context of bigger brains. An evolutionary compromise has been reached by the process, discussed elsewhere, of neoteny. This has provided a very successful thread – at least, until now. Brains remain circumscribed by their evolutionary origins. Alfred Sherwood Romer had a view that the brain had evolved in two parts, the 'primitive' and the 'modern'. The primitive brain was the home of existential intelligence, where functions such breathing, reproduction, excretion, intestinal movement and heartbeat, functions that are more than half a billion years old, were controlled without conscious intervention. The modern brain – cerebral hemispheres and associated structures – was where consciousness and behaviour were regulated, so it looked after everything else.

This does not, however, tell us much about the multifunctional capacity of the modern brain. Average men or women are equipped with a set of intelligences, which they can apply, individually or simultaneously, to any problem they encounter. Sadly, most of us have the different intelligences developed to different degrees and must do the best we can.

Studies have shown that infants in orphanages may die without emotional care before they are six years old. If they do not hear a language spoken before they are nine, they will never develop speech. If they do not use their hands (feet can replace them) as they grow, their thinking capacity will be limited. The conclusion is that humans are inherently social animals, co-dependent on opportunities for learning, and needing manual as well intellectual stimulation for growth. They learn to integrate with their social groups and learn of the rewards of cooperative behaviour.

As well as being generalists, humans – and to a lesser, extent bonobos and chimpanzees – are closely cooperative. In a social group all the ways of thinking may not need to be represented in all members but, ideally, the whole spectrum should be employed when addressing an intractable problem. There are benefits from working together. Life experiences hone different intelligences in different ways. Individual collective thinking is perhaps possible for gifted polymaths, but collaborative thinking is available within any group of human beings. If cells cooperate, why not minds?

In evolutionary terms, one must assume that animals are equipped with different subsets of intelligences, unknown to us, to enable them to solve the problems posed by their habits of life. They may possess intelligences that humans cannot experience, such as the capacity to perceive and interpret magnetic fields. An eagle, whose visual acuity is essential to its lifestyle, will possess an intelligence consequent on this, while a dog, whose world is largely an olfactory one, will have at least one of its intelligences predicated on an exquisite sense of smell. Humans, being the supreme generalists, are gifted with the wide selection of intelligence and ways of thinking, but blunter sense organs.

Constructing the human contribution to Gaia is a continuing process. At present, in the so-called Anthropocene, we are fixated on the effects of climate change, and especially the human contribution to it by the burning of fossil fuels. We can see extinction happening all around us, our world is changing as we look at it. The ice caps at the poles and in Greenland are melting, the sea is warming, sea levels are rising and weather patterns are changing. We should be ashamed of our part in it. We are at last becoming concerned because our coastal cities, where a large proportion of humanity resides, are threatened. But in spite of all these disruptions, Gaia, this planetary homeostatic mechanism, rolls on, not in the slightest bit concerned, towards a new steady state.

Major evolutionary changes – those changes that end up altering the shapes of animals and plants – are difficult for us to see in our short lifetimes. What we can detect are the myriad small changes that together may have a large effect at some time in the future. And sometimes we can use our new genome technology to look into the past and calculate how long it was since a change to the DNA of a particular population of organisms occurred. A whole meeting of the American Association of Science was devoted to the Biology of Genomes in 2016. Its findings were reported by Elizabeth Pennisi (2016). Here are some more of the results of several of the studies on humans.

Evolutionary changes can be detected in melanin regulation (skin colour, vitamin D, eye colour). Over the last few thousand years in women, hip size has been increasing, together with head circumference in babies, suggesting that there is continued modification of the neotenic changes that gave rise to modern humans in the first place. Genes for lactose tolerance have also spread rapidly through the British population in the last 2,000 years, presumably reflecting the historical growth of dairy farming. The introduction of milk in the diet had at least one unfortunate repercussion – the spread of tuberculosis (consumption) in Victorian England. In 1924 free milk (now tuberculin tested) in schools was introduced, and so a change of behaviours and increased health followed the first impact of the incorporation of a new component to the human diet; social evolution at its best.

These examples are not, by a long chalk, a summary of all that is happening to the human genome. We belong to an order of which we are the only extant members, yet it is quite clear that we are showing considerable variation, under the impact of our rapidly changing environment. But suppose we consider an order, Coleoptera (beetles), say, with 350,000 described species (or about 22 per cent of all described organisms), or the phylum Nematoda (roundworms), of which 22,000 species have been described, and there may yet be as many 80 million still to go. Now multiply those figures by the number of individuals there are in each species and the brain begins to spin. Yet each one of those individuals is interacting with its changing environment – and, itself, is changing little by little. Evolution is a long-term, widespread continuous process, driven by the process of natural selection and, in humans, social selection.

There has never been a better time to tell this story of an interconnected planet. From an evolutionary point of view, although some of the threads may get broken, change is an opportunity to spin new ones. A vacated peak in the adaptive landscape inevitably becomes occupied by an existing species from elsewhere. If the occupation is successful, its descendants radiate throughout the void, changing gradually as they do so. This is called adaptive radiation and is an important driver of evolution.

It happens every day. As they have little competition, great white sharks seem to be extending their range into the once too-cold waters around the south of Australia. What new opportunities are being extended to their offspring? The bleaching of the Great Barrier Reef provides niches for other plants and animals that may be less attractive to our eyes, but our aesthetic sensibilities don't overlap with other species in the overall scheme of life. Plants and animals are advancing up mountains as they become warmer. We do not live long enough to fully perceive these changes and whether we like them is irrelevant. We still have to deal with them.

The environment has moulded humans on their journey to the present. In the last 10,000 years, however, we have gone beyond a passive response to environmental change and have altered and are altering, at an increasing rate, the connections that have helped to create the steady state of Gaia. Notably, this has happened by increasing the return of carbon to the atmosphere in the form carbon dioxide, releasing it from the ancient coal measures at a rate so great that Gaia is unable to sustain the stable atmospheric dynamics that obtained in the middle of the eighteenth century. For those who have eyes to see, this is made manifest by rising global temperatures and atmospheric perturbation in the form of more extreme weather. Gaia is fighting back, quite impersonally, as all homeostatic systems do. We may end up as 'collateral damage'. The more pessimistic students of evolution are suggesting that the effects exerted by humans on the environment have brought us to the end of our line. Others, no less pessimistic suggest that, as humans are innately competitive, they will bring themselves to an end in ever-escalating wars over natural resources, like fresh water and arable land, that might become scarce in Gaia's new steady state.

This book has taken a different view. Throughout, it has argued that the whole evolutionary trajectory is built on cooperation more than competition. The metaphors and other comparisons we have used to emphasise the link between the human condition and the biosphere demonstrate this, we hope, even if some are rather fanciful. And now we are at the point where we have to rely on the major evolutionary attributes of an otherwise generalist species. Humans are quite good at a lot of things but hopelessly outplayed on a one to one comparison; we do not have an eagle's acuity of vision, a dog's olfactory discrimination, a bat's range of hearing, a pigeon's homing instinct, an elephant's strength, a horse's fleetness of foot, a gibbon's gymnastic ability – you can add to this list yourself. What we do have is a larger brain and manipulative hands. We can imagine ourselves doing what other organisms do and so we have made telescopes to enhance out visual acuity; gas chromatographs so that we compete with the olfaction of our pet dog; global positioning systems so that we can travel without being able to sense the Earth's magnetic field; cranes, whose strength is far greater than elephants; cars that outrun horses; climbing apparatus that will leave gibbons far behind.

In the previous pages, we have looked at the popular perceptions of 'nature, red in tooth and claw' and 'survival of the fittest' in a different way, one we think that is more consistent with Darwin's own view of life. A literal reading of the first of these applies only to vertebrate carnivores and, given that vertebrates make up a tiny fraction of the less that 1 per cent of biomass of animals on Earth, it is insignificant as a way of life. Invertebrates feed on other invertebrates but this is still only a fraction of this 1 per cent, the rest of whom are herbivores. Yet carnivores have their uses: returning wolves to Yellowstone National Park rejuvenated a tired and depleted ecosystem. Similar observations about the roles of top carnivores have been made elsewhere and now there is a movement to return wolves to Europe. We must consider our own place in this – but it does not mean that we must consent to be eaten!

What would the plants, by far the greatest proportion of the biosphere, make of it all? We would not go so far as to say they consent to being eaten – except those parts such as fruits, seeds and nuts, that are 'designed' to be an attractive food source for animals – and there are plenty of plant adaptations that discourage potential grazers and browsers. They are, however, so constructed that, generally, they survive losing a few leaves and twigs to a marauding giraffe. In so doing they help to nurture the giraffe – and all the inhabitants of its large rumen, not to mention the ultimate benefits of its dung to pasture. Size is just another characteristic of organisms and does not necessarily warrant greater consideration than any other. Looked at in this way, the giraffe, as a dung producer, is simply the first step of a huge matrix of interdependences.

'Survival of the fittest' always conjures up in the popular mind the idea of competition. And that usually means, to the average farmer or gardener, competition between their crop plants and invading weed species. There is a famous cartoon from *Punch* of a beautiful cottage garden and a plaque on the cottage wall quoting TE Brown's 'A garden is a lovesome thing, God wot'. The proud gardener is leaning on his wall and remarking to an admirer 'You shoulda seen the mess it was in when God had it to hisself!' A well-kept cottage garden (or farm) is so far out of natural equilibrium that it is indeed a battlefield requiring constant supportive assaults from gardener or farmer.

'Survival of the fittest' might be better rephrased as the 'survival of the luckiest'. Had this been understood in the nineteenth century perhaps a whole lot of human misery might have been avoided, because natural selection really is about luck. And that luck is going for an organism that finds itself in the right place in the wider ecosystem, a place that provides an opportunity for growth and reproduction. As we have seen, it pays sometimes to be a hyper-specialist, like insects with a very specific food range, such as the cabbage white butterfly on brassicas. The butterflies do well as long as some disease does not wipe out cabbages and their kin. For humans, concentrating on a single staple food is likewise dangerous, as those caught up in the tragedy of the Irish famine in the nineteenth century found. That was a special case engineered by cynical politicians. We are generalists; we, like pigs, can eat almost anything, including each other. Normally, if something wipes out potatoes or acorns, we just move on to something else.

As the environment constantly changes it is the task of every living thing to track those changes and adapt to them. Some are always going to be luckier than others that find their favoured environment has become too hot, too cold, too wet, too dry, too crowded, exhausted of nutrients and so on. Nothing is lucky all the time; extinction is commonplace. Something will push the hopeful evolver over the brink, even if it takes worldwide volcanic activity or an asteroid. We are here, and our lineage, from the first sign of life, is exactly as long as every other living thing. We all have a 4-billion-year history of successful adaptation. Together we form the biosphere, the network of biological creation that is the engine of Gaia. We also form our own human net, that Teilhard de Chardin called the noosphere and C Wright Mills 'the sociological imagination'. It is our physical link with the past and fuels our creative capacity for our future. We have enhanced it with a communications technology that we are still learning to control. What a marvellous past it has been and what a marvellous future we may yet have. Sir Charles Sherrington, a neurobiologist who was awarded the Nobel Prize in 1932 would certainly agree. In 1951 he wrote that his book, *Man on his Nature*,

> stresses the view that man [sic] is a product, like so much else, of the play of natural forces acting on the material and under the conditions past and present obtaining on the surface of our planet ... from the side [of the planet] arising shape after shape past fancy. And latterly among them some imbued with sense and thought ... now yielding thoughts and values.

Recently, a paper entitled 'Gaia 2.0', by Lenton and Latour (2018), put forward a plausible mechanism by which Gaia herself might evolve. It is derived from observations on automata that reset, or 'reboot' themselves. Each time they reboot, they tend to move to a condition of greater stability. Gaia has suffered half a dozen 'great' extinctions – or reboots – and 20-odd 'lesser' extinctions, in each case leading to a new period of stability during which complexity appears to have increased. Thus, each reboot is a resetting, as long as it is not a total extinction event, and Gaia can build on what has gone before. The evolutionary tendency is thus towards stability. Based on this, the so-called Anthropocene is merely a harbinger of a new steady state.

Ancient cells have *always* had a role and have *always* been with us in one form or another, inside and out. The first great extinction was brought about as a consequence of one group of ancient cells discovering photosynthesis and poisoning the Earth with oxygen. The response was the emergence of the modern cell that combined the activities of several ancient cells. Ancient cells, the ones that did not enter into such arrangements, persist in astronomical numbers to this day. They join with plants, animals and fungi to contribute to the integrity of Gaia. They, in the form of humanity, have created intelligence. Gaia, Darwin's 'tangled bank' writ large, is thus a testament to the ultimate triumph of ancient cells whose origins go back almost to the origin of the Earth.

Since that time, there have been many steady states of Gaia, all of them capable of supporting life in one form or another. Sometimes the shift to a new steady state has been gradual, sometimes it has been as a result of catastrophe. Until the Industrial Revolution, the steady state shifted imperceptibly. Now we can perceive it happening and, what's more, understand our own less than glorious – perhaps even catastrophic – contribution. If humans are an integral part of Gaia, it is reasonable to assume that human consciousness, an essential attribute of humanity, has a role to play in establishing the new steady state.

If we don't play our part, we are just another catastrophe.

BIBLIOGRAPHY

Adams, D. 1979. *The Hitchhiker's Guide to the Galaxy*. Pan Books.

Aiello, LC. 2010. Five Years of *Homo Floresiensis*. *Am J Phys Anthropol* **142**. 167–179.

Alexander, CF. 1848. 'All Things Bright and Beautiful'. In: *Hymns for Little Children*. Available at: hymnary.org/text/all_things_bright_and_beautiful.

Alvarez, LW, W Alvarez, F Asaro and HV Michel. 1980. Extraterrestrial Cause for the Cretaceous-Tertiary Extinction. *Science* **208**. 1095–1108. doi.org/10.1126/science.208.4448.1095.

Andrewartha, HG and LC Birch. 1954. *The Distribution and Abundance of Animals*. University of Chicago Press.

Aplin, LM, BC Sheldon and J Morand-Ferron. 2013. Milk Bottles Revisited: Social Learning and Individual Variation in the Blue Tit, *Cyanistes Caeruleus*. *Anim Behav* **85**. 1225–1232. doi.org/10.1016/j.anbehav.2013.03.009.

Argue, D, CP Groves, MSY Lee and WL Jungers. 2017. The Affinities of *Homo Floresiensis* Based on Phylogenetic Analyses of Cranial, Dental, and Postcranial Characters. *J Hum Evol* **107**. 107–133. doi.org/10.1016/j.jhevol.2017.02.006.

Asimov, I. 1954. *The Caves of Steel*. Doubleday.

Ayala, FJ and JJ Cela-Conde. 2017. *Processes in Human Evolution: The Journey from Early Hominins to Neanderthals and Modern Humans*. Oxford University Press. doi.org/10.1093/oso/9780198739906.003.0011.

Baker, CJ and CJ Orlandi. 1995. Active Oxygen in Plant Pathogenesis. *Ann Rev Phytopath* **33**. 299–321.

Balter, M. 2013. How to Survive a Siberian Winter. *Science*, January 28. Available at: www.sciencemag.org/news/2013/01/how-survive-siberian-winter.

Bandi, C, B Slatko and SI O'Neill. 1999. *Wolbachia* Genomes and the Many Faces of Symbiosis. *Parasitol Today* **15**. 428–429. doi.org/10.1016/s0169-4758(99)01543-4.

Barrass, C. 2019. Your Mystery Microbiome. *New Scientist* **242**. 28–31.

Barthes, R and L Duisit. 1975. An Introduction to the Structural Analysis of Narrative. *New Lit Hist* **6**. 237–272.

Beard, C. 2004. *The Hunt for the Dawn Monkey.* University of California Press.

Belkin, S, DC Nelson and HW Jannasch. 1986. Symbiotic Assimilation of CO_2 in Two Hydrothermal Vent Animals, the Mussel *Bathymodus Thermophilus* and the Tube Worm *Riftia Pachyptila. Biol Bull* **170**. 110–121. doi.org/10.2307/1541384.

Berrill, NJ. 1955. *The Origin of Vertebrates.* Clarendon Press, Oxford.

Blake, W. 1863. *Auguries of Innocence.* Available at: www.poetryfoundation.org/poems/43650/auguries-of-innocence.

Bolk, L. 1929. Origin of racial characteristics in man. *American Journal of Physical Anthropology* **13** (1). 1–28. doi.org/10.1002/ajpa.1330130123.

Bonner, JT. 2010. *The Social Amoebae: The Biology of Cellular Slime Moulds.* Princeton University Press.

Bordenstein, SH and KR Theis. 2015. Host Biology in the Light of the Microbiome: Ten Principles of Holobionts and Hologenomes. *Plos Biol* **13**. e1002226. doi.org/10.1371/journal.pbio.1002226.

Brahic, C. 2018. Traces of Mystery Ancient Humans Lurking in our Genomes. *New Scientist*, October 13. Available at: www.newscientist.com/article/mg24031992-600-traces-of-mystery-ancient-humans-found-lurking-in-our-genomes/.

Brockman, J, ed. 1996. *The Third Culture: Beyond the Scientific Revolution.* Simon and Schuster.

Brown, VA. 2008. *Leonardo's Vision.* Sense Publishers.

Brown, VA and JA Harris. 2014. *The Human Capacity for Transformational Change.* Routledge.

Browne, J. 2003. *Charles Darwin: Vol. 1 Voyaging* and *Vol. 2 The Power of Place.* Pimlico Press.

Bryant, C. 1983. Intraspecies Variations of Energy Metabolism in Parasitic Helminths. *Int J Parasit* **13**. 327–332.

Burt, C. 1909. Experimental Tests of General Intelligence. *Brit J Psychol* **3**. 94–177.

Byatt, AS. 2012. *Ragnarok: The End of the Gods*. Canongate Books.

Cann, RL, M Stoneking and AC Wilson. 1987. Mitochondrial DNA and Human Evolution. *Nature* **325**. 31–36. doi.org/10.1038/325031a0.

Carbonell, E, M Mosquera and CR Pale. 2006. The Emergence of a Symbolic Behaviour: The Sepulchral Pit of Sima de los Huesos, Sierra de Atapuerca, Burgos, Spain. *Comptes Rendus Palevol* **5**. 155–160. doi.org/10.1016/j.crpv. 2005.11.010.

Cardona, A and 16 others. 2014. Genome-Wide Analysis of Cold Adaptation in Indigenous Siberian Populations. *PLoS One* **9**. e98076.

Carroll, L. 1871. *Through the Looking-Glass*. Macmillan and Co., London. Available at: www.gutenberg.org/files/12/12-h/12-h.htm.

Check, E. 2002. Worm Cast in Starring Role for Nobel Prize. *Nature* **49**. 548. doi.org/10.1038/419548a.

Checkland, P. 1999. *Systems Thinking, Systems Practice: Includes a 30 Year Retrospective*. Wiley.

Chief Seattle. 1854. *Letter to All*. Available at: www.csun.edu/~vcpsy00h/seattle. htm.

Chouard, T. 2010. Evolution: The Revenge of the Hopeful Monster. *Nature* **463**. 864–867. doi.org/10.1038/463864a.

Christensen, C. 2003. *The Innovator's Dilemma*. Collins.

Clarkson, C and 28 others. 2017. Human Occupation of Northern Australia by 65,000 Years Ago. *Nature* **547**. 306–310. doi.org/10.1038/nature22968.

Corning, PA. 2002. The Emergence of 'Emergence': Now What? *Emergence*. **4**. 54–71. doi.org/10.1207/s15327000em0403-05.

Darwin, C. 1851. *Geological Observations on Coral Reefs, Volcanic Islands and on South America*. Smith, Elder & Co., London. Available at: darwin-online.org. uk/converted/pdf/1851_Geology_F274.pdf.

Darwin, C. 1859. *On the Origin of Species by Means of Natural Selection, or the Preservation of Favoured Races in the Struggle for Life.* John Murray, London. Available at: darwin-online.org.uk/content/frameset?itemID=F373&viewtype=text&pageseq=1. doi.org/10.5962/bhl.title.68064.

Darwin, C. 1871. *The Descent of Man, and Selection in Relation to Sex.* John Murray, London. Available at: darwin-online.org.uk/EditorialIntroductions/Freeman_TheDescentofMan.html. doi.org/10.5962/bhl.title.24784.

Darwin, C. 1872. *The Expression of the Emotions in Man and Animals.* John Murray, London. Available at: darwin-online.org.uk/EditorialIntroductions/Freeman_TheExpressionoftheEmotions.html.

Darwin, C. 1881. *The Formation of Vegetable Mould Through the Action of Worms, with Observations on their Habits.* John Murray, London. Available at: darwin-online.org.uk/content/frameset?itemID=F1357&viewtype=text&pageseq=1. doi.org/10.5962/bhl.title.121748.

Darwin, F, ed. 1887. *The Life and Letters of Charles Darwin, Including an Autobiographical Chapter.* Vol. 1. John Murray, London. Available at: darwin-online.org.uk/content/frameset?itemID=F1452.1&viewtype=text&pageseq=1. doi.org/10.5962/bhl.title.50683.

Dawkins, R. 1996. *Climbing Mount Improbable.* Viking.

De Bono, E. 1999. *Six Thinking Hats.* Back Bay Books.

De Waal, F. 2014. *The Bonobo and the Atheist.* W.W. Norton and Co.

Dennett, DC. 2018. *From Bacteria to Bach and Back.* W.W. Norton and Co.

Desmond, A. 1989. *The Politics of Evolution.* University of Chicago Press.

Desmond, A and J Moore. 1991. *Darwin.* Michael Joseph, London.

Diamond, J. 1992. *The Third Chimpanzee.* Harper Collins.

Dobzhansky, T. 1973. Nothing in Biology Makes Sense except in the Light of Evolution. *Am Biol Teach* **35**. 125–129. doi.org/10.2307/4444260.

Dover, G. 2000. *Dear Mr Darwin: Letters on the Evolution of Life and Human Nature.* Weidenfeld and Nicholson. London.

Durston, AJ. 2013. Dictyostelium: The Mathematician's Organism. *Curr Genomics* **14**. 255–360.

Edelson, E. 1999. *Gregor Mendel and the Roots of Genetics.* Oxford University Press.

Eggleton, T. 2012. *A Short Introduction to Climate Change.* Cambridge University Press.

Eldredge, N and SJ Gould. 1972. Punctuated Equilibria: An Alternative to Phyletic Gradualism. In: TJM Schopf, ed., *Models in Paleobiology.* Freeman, Cooper and Co., San Francisco. doi.org/10.5531/sd.paleo.7.

Eman, MY, HS Nahla and IAM Eissa. 2014. Parasitological Studies of the Isopoda (Cymothoidae) Parasites Infesting some Marine Fishes at Suez Cana Area of Ismailia Province Egypt, with a Key to the Cymathoid Genera. *Life Sci J* **11**. 227–231.

Evans, AD and K Lee. 2013. Emergence of Lying in Very Young Children. *Dev Psych* **49**. 1958–1963.

Fagan, D. 2017. *Wake Up: The Nine Hashtags of Digital Disruption.* University of Queensland Press.

Fitt, WK. 1984. The Role of Chemosensory Behaviour of *Symbiodinium Microadriaticum*, Intermediate Hosts, and Host Behaviour in the Infection of Coelenterates and Molluscs and with Zooanthellae. *Mar Biol* **81**. 9–17. doi.org/10.1007/bf00397620.

Francis, RC. 2011. *Epigenesis: How Environment Shapes our Genes.* Norton and Co., New York.

Frost, R. 1914. *Mending Wall.* Available at: www.poetryfoundation.org/poems/44266/mending-wall.

Fruth, H. 2015. *The Flow of Money.* Policom Corporation.

Gardner, H. 1983. *Frames of Mind.* Basic Books.

Gardner, H. 2007. *Five Minds for the Future.* Harvard Business School Press.

Gilman, C. 1890. *Similar Cases.* Available at: www.poetryfoundation.org/poems/46060/similar-cases.

Gleick, J. 1987. *Chaos.* Heinemann, London.

Goldschmidt, RC. 1940. *The Material Basis of Evolution.* Yale University Press.

Gordon, D. 2010. *Ant Encounters: Interaction Networks and Colony Behaviour.* Princeton University Press.

Grossman, L. 2017. Nuclear Holiday. *New Scientist* **234**. 20–21.

Hanschen, ER and 23 others. 2016. The *Gonium Pectorale* Genome Demonstrates the Co-Option of Cell-Cycle Regulation During the Evolution of Multicellularity. *Nat Commun* **7**. Available at: www.nature.com/articles/ ncomms11370. doi.org/10.1038/ncomms11370.

Harris MP, SM Hasso, WJ Ferguson and JF Fallon. 2006. The Development of Archosaurian First Generation Teeth in a Chicken Mutant. *Curr Biol* **16**. 371–377. doi.org/10.1016/j.cub.2005.12.047.

Havird, JC, MD Hall and DK Dowling. 2015. The Evolution of Sex: A New Hypothesis Based on Mitochondrial Mutational Erosion. *Bioessays.* **37**. 951–958. doi.org/10.1002/bies.201500057.

Heylighen, F. 1998. What Makes a Meme Successful: Selection Criteria for Cultural Evolution. *Proc. 15th Int. Congress on Cybernetics*. Association Internat de Cybernétique, Namur. 418–423.

Horne, D. 1970. *God is an Englishman.* TBS Ltd.

Hublin, JJ and 10 others. 2017. New fossils from Jebel Irhoud, Morocco and the Pan-African origin of *Homo Sapiens. Nature* **546**. 289–292. doi.org/10.1038/ nature22336.

Huxley, A. 1932. *Brave New World.* Chatto and Windus.

Huxley, J. 1942. *Evolution: The Modern Synthesis.* George Allen and Unwin, London.

Jablonka, E and M Lamb. 2014. *Evolution in Four Dimensions.* MIT Press.

Jami, E and I Mizrahi. 2012. Composition and Similarity of Bovine Rumen Microbiota across Individual Animals. *Plos One* **7**. e33306. doi.org/10.1371/ journal.pone.0033306.

John, DT. 1982. Primary Amebic Meningoencephalitis and the Biology of *Naegleria Fowleri. Ann Rev Microbiol* **36**. 101–123. doi.org/10.1146/annurev. mi.36.100182.000533.

Johnson, ED. 1926. A Comparison of the Juvenile and Adult Leaves of *Eucalyptus Globulus. New Phytol* **25**. 202–212. doi.org/10.1111/j.1469-8137.1926.tb06691.x.

Kanther, M and JF Rawls. 2010. Host–Microbe Interactions in Developing Zebra Fish. *Curr Opin Immunol* **22**. 10–19.

Kauffman, S. 1995. *At Home in the Universe: The Search for the Laws of Complexity.* Penguin Books.

Keller, EF. 1983. *A Feeling for the Organism: The Life and Work of Barbara McClintock*. Henry Holt and Company.

Kerney, R. 2011. Symbioses Between Salamander Embryos and Green Algae. *Symbiosis* **54**. 107–117. doi.org/10.1007/s13199-011-0134-2.

Kipling, R. 1911. *The Female of the Species*. Available at: www.potw.org/archive/potw96.html.

Koestler, A. 1971. *The Case of the Midwife Toad*. Hutchinson &Co, London.

Kropotkin, PA. 1902. *Mutual Aid; A Factor in Evolution*. McClure, Philips & Co. Available at: www.gutenberg.org/ebooks/4341.

Kurzweil, R. 2016. *The Singularity is Near*. Penguin Books, London.

Lafferty, KD and AM Kuris. 2009. Parasitic Castration: The Evolution and Ecology of Body Snatchers. *Trends Parasitol* **25**. 564–572. doi.org/10.1016/j.pt.2009.09.003.

Laland, K. 2016. Evolution Evolves: Beyond the Selfish Gene. *New Scientist*, September 21. Available at: www.newscientist.com/article/mg23130920-600-evolution-evolves-beyond-the-selfish-gene/. doi.org/10.1016/s0262-4079(16)31748-1.

Laranjeiras, CC and SIC Portela. 2016. The Carnot Cycle and the Teaching of Thermodynamics: A Historical Approach. *Physics Ed* **51**. 055013. doi.org/10.1088/0031-9120/51/5/055013.

Lawson, H. 1922. *On the Night Train*. Available at: www.poetryverse.com/henry-lawson-poems/on-the-night-train.

Legg, S and M Hutter. 2007. A Collection of Definitions of Intelligence. *Frontiers in Artificial Intelligence and Applications*. **157**. 17–24. Available at: arxiv.org/abs/0706.3639.

Lenton, TM and B Latour. 2018. Gaia 2.0. *Science* **361**. 1066–1068.

Lesk, AM. 2017. *Introduction to Genomics*. 3rd ed. Oxford University Press.

Lofting, H. 1926. *Dr Dolittle's Zoo*. Jonathan Cape.

Lovelock, J. 2013. *Homage to Gaia*. Souvenir Press.

Lovelock, J and S Epton. 1975. The Quest for Gaia. *New Scientist* **65**. 304–306.

Malthus, TR. 1798. *An Essay on the Principle of Population*. J. Johnson in St Pauls Churchyard, London. Available at: www.esp.org/books/malthus/population/malthus.pdf. doi.org/10.1007/978-1-349-81729-0.

Mann, A. 2010. Sponge Genome Goes Deep. *Nature* **466**. 673. doi.org/10.1038/466673a.

Margulis, L. 1970. *The Origin of Eukaryotic Cells.* Yale University Press.

Margulis, L, ed. 1991. *Symbiosis as a Source of Evolutionary Innovation: Speciation and Morphogenesis.* MIT Press.

Margulis, L and D Sagan. 1997. *Microcosmos: Four Billion Years of Microbial Evolution.* University of California Press.

Massey, C. 2017. *Call of the Reed Warbler: A New Agriculture – a New Earth.* University of Queensland Press.

McCutcheon, JP and CD von Dohlen. 2011. An Interdependent Metabolic Patchwork in the Nested Symbiosis of Mealybugs. *Curr Biol* **21**. 1366–1372. doi.org/10.1016/j.cub.2011.06.051.

McCutcheon, JP, BR McDonald and NA Moran. 2009. Convergent Evolution of Metabolic Roles in Bacterial Co-Symbionts of Insects. *PNAS* **106**. 15394–15399. doi.org/10.1073/pnas.0906424106.

McLuhan, M. 1967. *The Medium is the Message.* Random House.

Mead, M. 1928. *Coming of Age in Samoa.* William Morrow & Co., New York.

Meadows, DH, DL Meadows, J Randers and WW Behrens. 1974. *The Limits to Growth: A Report for the Club of Rome's Project on the Predicament of Mankind.* Universe Books.

Meadows, DH, J Randers and DL Meadows. 2013. *Limits to Growth: The 30-year Update.* Chelsea Green Publishing.

Midgley, M. 1980. *Beast and Man.* Methuen & Co.

Midgley, M. 2004. *The Myths We Live By.* Routledge, Oxford.

Miller, H. 1919. *The Old Red Sandstone.* Everyman's Library (first published in 1851).

Mithen, S. 2005. *The Singing Neanderthals: The Origins of Music, Language, Mind and Body.* Harvard University Press.

Moore, JR and DJ Varricchio. 2016. The Evolution of Diapsid Reproductive Strategy with Inferences about Extinct Taxa. *PLoS ONE* **11**. e0158496. doi.org/10.1371/journal.pone.0158496.

Moran, NA and DB Sloan. 2015. Hologenome Concept: Helpful or Hollow? *PLOS Biol* **13**. e1002311. doi.org/10.1371/journal.pbio.1002311.

Mujer, CV, DL Andrews, JR Manhart and ME Rumpho. 1996. Chloroplast Genes Are Expressed During Intracellular Symbiotic Association of *Vaucheria Litorea* Plastids with the Sea Slug *Elysia Chlorotica*. *PNAS* **93**. 12333–12338. doi.org/10.1073/pnas.93.22.12333.

Mumford, L. 1961. *The City in History.* Harcourt Inc.

Nash, TH. 2008. *Lichen Biology.* Cambridge University Press.

Noble, D. 2006. *The Music of Life: Biology Beyond the Genes.* Oxford University Press.

Odum, EP. 1953. *Fundamentals of Ecology.* W.B. Saunders, Philadelphia.

Okamoto, N and GI McFadden. 2008. The Mother of All Parasites. *Future Microbiol* **3**. doi.org/10.2217/17460913.3.4.391.

Ostrom, E. 2000. Collective Action and the Evolution of Social Norms. *J Econ Perspect* **14**. 137–158.

Pascoe, B. 2018. *Dark Emu.* Magabala Books.

Pennisi, E. 2016. Humans are Still Evolving and We Can Watch it Happen. *Science,* May 17. Available at: www.sciencemag.org/news/2016/05/humans-are-still-evolving-and-we-can-watch-it-happen. doi.org/10.1126/science.aaf5727.

Pianka, ER. 1970. On r- and K- Selection. *Amer Natural* **104**. 592.

Power, DA, RA Watson, E Szathmary and R Mills. 2015. What Can Ecosystems Learn? Expanding Evolutionary Ecology with Learning Theory. *Biol Direct* **10**. Available at: biologydirect.biomedcentral.com/articles/10.1186/s13062-015-0094-1. doi.org/10.1186/s13062-015-0094-1.

Poznik, GD and 10 others. 2013. Sequencing Y Chromosomes Resolves Discrepancy in Time to Common Ancestor of Males versus Females. *Science* **341**. 562–565. doi.org/10.1126/science.1237619.

Pradeu, T. 2012. *The Limits of the Self: Immunology and Biological Identity.* Elizabeth Vitanza (transl.). Oxford University Press.

Quammen, D. 1996. *The Song of the Dodo.* Hutchinson, London.

Raper, KB. 1935. *Dictyostelium Discoideum*, a New Species of Slime Mold from Decaying Forest Leaves. *J Agric Res* **50**. 135–147.

Ratcliff, WC, JD Fankhauser, DW Rogers, D Greig and M Travisano. 2015. Origins of Multicellular Evolvability in Snowflake Yeast. *Nat Commun* **6**. Available at: pubmed.ncbi.nlm.nih.gov/25600558/. doi.org/10.1038/ncomms7102.

Rendu, W and 13 others. 2014. Evidence Supporting an Intentional Neandertal Burial at La-Chapelle-aux-Saints. *PNAS* **111**. 81–86.

Rittel, H and M Webber. 1973. Dilemmas in a General Theory of Planning. Reprinted in: N Cross, ed., *Developments in Design Methodology* (1984). J. Wiley & Sons.

Round, JL and SK Mazmanian. 2009. The Gut Microbiota Shapes Intestinal Immune Responses During Health and Disease. *Nat Rev Immunol* **9**. 313–323. doi.org/10.1038/nri2515.

Ruby, EG and KH Lee. 1998. The *Vibrio Fischeri – Euprymna Scolopes* Light Organ Association: Current Ecological Paradigms. *Appl Environ Microbiol* **1**. 805–812. Available at: www.ncbi.nlm.nih.gov/pmc/articles/PMC106330/. doi.org/10.1128/aem.64.3.805-812.1998.

Ruse, M. 2013. *The Gaia Hypothesis*. University of Chicago Press.

Sandler, L and E Novitski. 1957. Meiotic Drive as an Evolutionary Force. *Am Nat* **91**. 105–110. doi.org/10.1086/281969.

Savory, TH. 1936. *Mechanistic Biology and Animal Behaviour*. C.A. Watts and Co., London.

Scherer, KR. 2005. What Are Emotions? And How Can They be Measured? *Social Science Information* **44**. 695–729. doi.org/10.1177/0539018405058216.

Shea, BT. 1989. Heterochrony in Human Evolution. The Case for Neoteny Considered. *Am J Phys Anthropol* **32**. 69–101. doi.org/10.1002/ajpa.13303 20505.

Sherrington, C. 1951. *Man on his Nature*. Penguin Books Ltd.

Shubin, N. 2008. *Your Inner Fish*. Penguin.

Simmons, LW, M Beveridge, N Wedell and T Tregenza. 2006. Postcopulatory Inbreeding Avoidance by Female Crickets Only Revealed by Molecular Markers. *Molec Ecol* **15**. 3817–3824. doi.org/10.1111/j.1365-294X.2006. 03035.x.

Singer, C. 1959. *A History of Biology*. Abelard Schuman.

Smuts, JC. 1926. *Holism and Evolution*. MacMillan and Co. Ltd., London.

Soumana, IH, B Thicaya, G Simo and A Geiger. 2014. Comparative Gene Expression of *Wigglesworthia* Inhabiting Non-Infected and *Trypanosoma Brucei Gambiense*-Infected *Glossina Palpalis Gambiensis* Flies. *Front Microbiol* **5**. 620. doi.org/10.3389/fmicb.2014.00620.

Spengler, O. 1932 (reprint). *The Decline of the West* (2 volumes). Alfred A. Knopf.

Srivastava, M and 20 others. 2008. The *Tricoplax* Genome and the Nature of Placozoans. *Nature* **454**. 955–960. Available at: www.nature.com/articles/nature07191.

Stacey, PB and W Koenig, eds. 2006. *Cooperative Breeding in Birds.* Cambridge University Press.

Steele, EJ. 1981. *Somatic Selection and Adaptive Evolution: On the Inheritance of Acquired Characters.* University of Chicago Press.

Steele, EJ, RA Lindley and RV Blanden. 1998. *Lamarck's Signature.* Basic Books, New York.

Sutherland, JL. 1933. Protozoa from Australian Termites. *Quart J Micr Sci* **76**. 145–173.

Swift, J. 1733. *On Poetry: A Rhapsody.* Available at: www.online-literature.com/swift/3515/.

Teilhard de Chardin, P. 1959. *The Phenomenon of Man.* Harper.

Tennyson, Lord A. 1848. *In Memoriam A.H.H.* Available at: www.online-literature.com/tennyson/718/.

Tlaika, M, D Bebber, PR Darrah and SC Watkinson. 2008. Chapter 3 Mycelial Networks: Nutrient Uptake, Translocation and Role in Ecosystems. In: L Boddy, JC Frankland and P van West, eds, *Ecology of Saprotrophic Basidiomycetes.* British Mycological Society Symposia Series **28**. Academic Press. doi.org/10.1016/s0275-0287(08)80005-7.

Tomarev, SI, O Sundin, S Banerjee-Basu, MK Duncan, J-M Yang and J Platigorski. 1996. Chicken Homeobox Gene Prox 1 Related to *Drosophila Prospero* is Expressed in the Developing Lens and Retina. *Dev Dyn* **206**. 354–367. Available at: anatomypubs.onlinelibrary.wiley.com/doi/10.1002/%28SICI%291097-0177%28199608%29206%3A4%3C354%3A%3AAID-AJA2%3E3.0.CO%3B2-H. doi.org/10.1002/(sici)1097-0177(199608)206:4<354::aid-aja2>3.0.co;2-h.

Toynbee, A. 1972. *A Study of History: Abridged and Illustrated.* Oxford University Press, New York.

Tunnicliffe, V. 1991. The Biology of Hydrothermal Vents: Ecology and Evolution. *Oceanogr Mar Biol Ann Rev* **29**. 319–407.

Turchin, P. 2007. *War and Peace and War: The Rise and Fall of Empires.* Plume Books.

Van Den Berghe, PL and GM Mesher. 1980. Royal Incest and Inclusive Fitness. *Am Ethnol* **7**. 300–317. doi.org/10.1525/ae.1980.7.2.02a00050.

Van Valen, L. 1973. A New Evolutionary Law. *Evolutionary Theory* **1**. 1–30.

Visalberghi, E and DM Fragaszy. 1990. Food-washing Behaviour in Tufted Capuchin Monkeys, *Cebus Apella* and Crabeating Macaques, *Macaca Fascicularis*. *Anim Behav* **40**. 829–836. doi.org/10.1016/s0003-3472(05)80983-2.

Wani, ZA, T Ashraf, T Mohiuddin and SR Ul-Hassan. 2015. Plant-Endophyte Symbiosis, an Ecological Perspective. *Appl Microbiol Biotechnol* **99**. 2955–2965. doi.org/10.1007/s00253-015-6487-3.

Ward, P and J Kirschvink. 2015. *A New History of Life.* Bloomsbury Press.

Watson, JD and F Crick. 1953. Molecular Structure of Nucleic Acids: A Structure for Deoxyribose Nucleic Acid. *Nature* **171**. 737–738. doi.org/10.1038/171737a0.

Watson, RA and E Szathmary. 2016. How Can Evolution Learn? *Trends Ecol Evol* **31**. 147–157. doi.org/10.1016/j.tree.2015.11.009.

Wayman, E. 2012. Five Early Primates You Should Know. *Smithsonian Magazine*, October 13. Available at: www.smithsonianmag.com/science-nature/five-early-primates-you-should-know-102122862/.

Wiener, N. 1988. *The Human Use of Human Beings: Cybernetics and Society.* Martino Fine Books.

Wilson, AC and 11 others. 2010. Genomic Insight into the Amino Acid Relations of the Pea Aphid, *Acyrthsiphon Pisum*, with its Symbiotic Bacterium *Buchnera Aphidicola*. *Insect Mol Biol* **19**. 249–258. Special Issue: *The Aphid Genome.* doi.org/10.1111/j.1365-2583.2009.00942.x.

Winchester, S. 2004. *Krakatoa.* Penguin Books, London.

Yamaguchi, M and 9 others. 2012. Prokaryote or Eukaryote? A Unique Microrganism from the Deep Sea. *J Electron Micr* **61**. 423–431.

Yeats, WB. 1919. *The Second Coming.* Available at: www.poetryfoundation.org/poems/43290/the-second-coming.

Yeats, WB. 1933. *Among School Children.* Available at: www.poetryfoundation.org/poems/43293/among-school-children.

www.ingramcontent.com/pod-product-compliance
Lightning Source LLC
Chambersburg PA
CBHW042111220326
41599CB00048BA/7166